T0305805

Healthcare Industry 4.0

This book covers computer vision-based applications in digital healthcare industry 4.0, including different computer vision techniques, image classification, image segmentations, and object detection. Various application case studies from domains such as science, engineering, and social networking are introduced, along with their architecture and how they leverage various technologies, such as edge computing and cloud computing. It also covers applications of computer vision in tumor detection, cancer detection, combating COVID-19, and patient monitoring.

Features:

- Provides a state-of-the-art computer vision application in the digital healthcare industry.
- Reviews advances in computer vision and data science technologies for analyzing information on human function and disability.
- Includes practical implementation of computer vision application using recent tools and software.
- Explores computer vision-enabled medical/clinical data security in the cloud.
- Includes case studies from the leading computer vision integrated vendors, such as Amazon, Microsoft, IBM, and Google.

This book is aimed at researchers and graduate students in bioengineering, intelligent systems, and computer science and engineering.

Computational Intelligence Techniques
Series Editor: Vishal Jain

The objective of this series is to provide researchers with a platform to present state-of-the-art innovations, research, and design, and implement methodological and algorithmic solutions to data processing problems, designing and analyzing evolving trends in health informatics and computer-aided diagnosis. This series provides support and aid to researchers involved in designing decision support systems that will permit societal acceptance of ambient intelligence. The overall goal of this series is to present the latest snapshot of ongoing research as well as to shed further light on future directions in this area. The series presents novel technical studies as well as position and vision papers comprising hypothetical/speculative scenarios. It seeks to compile all aspects of computational intelligence techniques from fundamental principles to current advanced concepts. For this series, we invite researchers, academicians, and professionals to contribute, expressing their ideas and research in the application of intelligent techniques to the field of engineering in handbook, reference, or monograph volumes.

Computational Intelligence Techniques and Their Applications to Software Engineering Problems
Ankita Bansal, Abha Jain, Sarika Jain, Vishal Jain and Ankur Choudhary

Smart Computational Intelligence in Biomedical and Health Informatics
Amit Kumar Manocha, Mandeep Singh, Shruti Jain and Vishal Jain

Data Driven Decision Making using Analytics
Parul Gandhi, Surbhi Bhatia and Kapal Dev

Smart Computing and Self-Adaptive Systems
Simar Preet Singh, Arun Solanki, Anju Sharma, Zdzislaw Polkowski and Rajesh Kumar

Advancing Computational Intelligence Techniques for Security Systems Design
Uzzal Sharma, Parmanand Astya, Anupam Baliyan, Salah-ddine Krit, Vishal Jain and Mohammad Zubair Kha

Graph Learning and Network Science for Natural Language Processing
Edited by Muskan Garg, Amit Kumar Gupta and Rajesh Prasad

Computational Intelligence in Medical Decision Making and Diagnosis: Techniques and Applications
Edited by Sitendra Tamrakar, Shruti Bhargava Choubey and Abhishek Choubey

Applications of 5G and Beyond in Smart Cities
Edited by Ambar Bajpai and Arun Balodi

Healthcare Industry 4.0: Computer Vision-Aided Data Analytics
Edited by P. Karthikeyan, Polinpapilinho F. Katina and R. Rajagopal

For more information about this series, please visit: www.routledge.com/Computational-Intelligence-Techniques/book-series/CIT.

Healthcare Industry 4.0
Computer Vision-Aided Data Analytics

Edited by
P. Karthikeyan, Polinpapilinho F. Katina and
R. Rajagopal

CRC Press
Taylor & Francis Group
Boca Raton London New York

CRC Press is an imprint of the
Taylor & Francis Group, an **informa** business

Designed cover image: © Shutterstock

First edition published 2024
by CRC Press
2385 NW Executive Center Drive, Suite 320, Boca Raton FL 33431

and by CRC Press
4 Park Square, Milton Park, Abingdon, Oxon, OX14 4RN

CRC Press is an imprint of Taylor & Francis Group, LLC

© 2024 selection and editorial matter, P. Karthikeyan, Polinpapilinho F. Katina and R. Rajagopal; individual chapters, the contributors

ISBN: 978-1-032-38515-0 (hbk)
ISBN: 978-1-032-38517-4 (pbk)
ISBN: 978-1-003-34541-1 (ebk)

DOI: 10.1201/9781003345411

Typeset in Times
by KnowledgeWorks Global Ltd.

Contents

*Mahesh Lokhande, Kalpanadevi D, Vandana Kate,
Arpan Kumar Tripathi, Prakash Bethapudi*

Preface

The healthcare industry is rapidly evolving with the advent of technology, and Industry 4.0 is poised to revolutionize how we approach healthcare data analytics. In this book, we explore the use of computer vision in healthcare data analytics, specifically focusing on how it can improve patient outcomes and streamline healthcare operations.

We provide an overview of Industry 4.0 and how it impacts healthcare. We then delve into the various applications of computer vision in healthcare, including medical imaging, telemedicine, and patient monitoring. We also discuss the challenges and limitations of using computer vision in healthcare and potential solutions.

This book is aimed at healthcare professionals, data scientists, and anyone interested in learning about the intersection of computer vision and healthcare. We hope this book will serve as a valuable resource for understanding the potential of computer vision in healthcare data analytics and inspire new ideas for innovation in the field.

Acknowledgments

We would like to express our sincere gratitude to the National Chung Cheng University, Taiwan; Alliance University, India; and the University of South Carolina Upstate, USA, for their support and guidance throughout the development of this book. We would also like to thank the contributors from various intuitions for their valuable contributions and insights. Their expertise and dedication have been instrumental in the success of this project. We are grateful for the opportunity to have worked with such a talented and dedicated group of individuals.

Contributors

Geetha A
Alliance College of Engineering and
   Design
Alliance University
Bangalore, Karnataka, India

Parshuram N. Arotale
Kamalnayan Bajaj Institute of
   Engineering and Technology,
Pune, Maharashtra, India

Surendiran B
NIT Puducherry
India

Prakash Bethapudi
Vignan Institute of Technology and
   Science
Hyderabad, Telangana,India

Trupti V. Bhandare
Kamalnayan Bajaj Institute of
   Engineering and Technology
Pune, Maharashtra, India

Rajendra Kumar Bharti
Bipin Tripathi Kumaon Institute of
   Technology
Dwarahat, Uttarakhand, India

Shashank D. Biradar
Kamalnayan Bajaj Institute of
   Engineering and Technology
Pune, Maharashtra, India

Prarthita Biswas
CERD, Adamas University
Kolkata, West Bengal, India

Om Prakash C
CMR University
Bangalore, Karnataka, India.

Ramya D
Sathyabama Institute of Science and
   Technology
Chennai, Tamilnadu, India

Kalpanadevi D
Kalasalingam Academy of Research
   and Education
Srivilliputhur, Tamilnadu, India

Shaheen H
Faculty of Computing & Engineering
UWL – RAK
UAE

Niha K
Vellore Institute of Technology
Vellore,Tamilnadu, India

Valarmathi K
Panimalar Engineering College
Chennai, Tamil Nadu, India

Premanand K. Kadbe
Kamalnayan Bajaj Institute of
   Engineering and Technology
Pune, Maharashtra, India

Vandana Kate
Acropolis Institute of Technology and
   Research
Indore, Madhya Pradesh, India

Mohiuddin Ali Khan
Jazan University
Jazan, Saudi Arabia

Ramakrishna Kolikipogu
JNTUH, Stanley College of Engineering
& Technology for Women
Hyderabad, Telangana, India

Mahesh Lokhande
Bharat Institute of Engineering and
    Technology
Hyderabad, Telangana, India

Rajadurai Narayanamurthy
Swiss School of Business Management
    (SSBM)
Geneva, Switzerland

Gayathiri P
Nirmala College for Women
Coimbatore, Tamil Nadu, India

Balasaheb H Patil
Kamalnayan Bajaj Institute of
    Engineering and Technology
Pune, Maharashtra, India

Rajagopal R
Alliance College of Engineering and
    Design
Alliance University
Bangalore, Karnataka, India

Sharmikha Sree R
Sri Sairam Engineering College
Chennai, Tamil Nadu, India

Kalpana R A
Sri Sairam Engineering College
Chennai, Tamil Nadu, India

Selvarani Rangasamy
Alliance College of Engineering and
    Design
Alliance University
Bangalore, Karnataka, India

Amutha S
Vellore Institute of Technology
Vellore, Tamil Nadu, India

Dinesh Kumar S
Sri Sairam Engineering College
Chennai, Tamil Nadu, India

Meera S
Sri Sairam Engineering College
Chennai, Tamil Nadu, India

Himanshu Sharma
J B Institute of Engineering and
    Technology
Hyderabad, Telangana, India

Chetan J. Shelke
Alliance College of Engineering and
    Design
Alliance University
Bangalore, Karnataka, India

Aaliyah Siddiqui
Symbiosis Centre for Management
    Studies
Nagpur, India

Mujahid Siddiqui
Dr. Ambedkar Institute of Management
    Studies and Research
Deekshabhoomi, Nagpur, India

Arpan Kumar Tripathi
Shri Shankaracharya
    Technical Campus
Bhilai, Chhattisgarh, India

Bhandare Trupti Vasantrao
Alliance College of Engineering and
    Design
Alliance University
Bangalore,Karnataka, India

Kamalnayan Bajaj Institute
    of Engineering and Technology
Pune, Maharashtra, India

About the Editors

P. Karthikeyan obtained his Bachelor of Engineering (B.E.) in Computer Science and Engineering from Anna University, Chennai, and Tamil Nadu, India, in 2005 and received his Master of Engineering (M.E.) in Computer Science and Engineering from Anna University, Coimbatore, India, in 2009. He completed his Ph.D. degree in ICE from the Anna University, Chennai, in 2018. He is currently completing postdoctoral research at National Chung Cheng University, Taiwan. He is skilled in developing projects and carrying out research in the area of cloud computing and deep learning with programming skills in Java, Python, R and C. He has published over 20 articles in international journals with high impact factors and has presented his work at more than 10 international conferences. He has served as a reviewer for prestigious publishers such as Elsevier, Springer, and Inderscience, as well as Scopus indexed journals. Additionally, he currently serves as an editorial board member for several journals, including the *EAI Endorsed Transactions on Energy Web*, *The International Arab Journal of Information Technology*, and the *Blue Eyes Intelligence Engineering and Sciences Publication Journal*.

Polinpapilinho F. Katina is an Assistant Professor in the Department of Informatics and Engineering Systems at the University of South Carolina Upstate, Spartanburg, South Carolina, USA. He previously served as a Postdoctoral Researcher for the National Centers for System of Systems Engineering in Norfolk, Virginia, USA, and Adjunct Professor in the Department of Engineering Management and Systems Engineering at Old Dominion University, Norfolk, Virginia, USA. Dr. Katina holds a B.Sc. in Engineering Technology with a minor in Engineering Management, a M. Eng. in Systems Engineering, and a Ph.D. in Engineering Management/Systems Engineering, all from Old Dominion University. He has received additional training from, among others, the Environmental Systems Research Institute in Redlands, California, USA, and Politecnico di Milano in Milan, Italy. His areas of research/teaching include, among others, complex system governance, critical infrastructure systems, decision-making and analysis (under uncertainty), emerging technologies, energy systems (smart grids), engineering management, infranomics, manufacturing systems, system of systems, systems engineering, systems pathology, and systems thinking.

R. Rajagopal received his Bachelor of Engineering (B.E.) in Computer Science and Engineering from Anna University, Chennai, and Tamil Nadu, India, in 2005, his Master of Engineering (M.E.) in Computer Science and Engineering from Anna University, Coimbatore, India, in 2009, and his Ph.D. degree from Anna University, Chennai, in 2018. He has served as a reviewer for reputed Scopus indexed journals and Inderscience. He is also an editorial board member of several journals, including the *International Journal of Mathematical Science, Engineering, and Sciences Publication*. His research work has been published in over 22 international journals, and he has presented at more than 20 international conferences.

1 Introduction to Computer Vision-Aided Data Analytics in Healthcare Industry 4.0

Ramakrishna Kolikipogu
Ramya D
Mujahid Siddiqui
Aaliyah Siddiqui
Om Prakash C

CONTENTS

DOI: 10.1201/9781003345411-1

1.1 INTRODUCTION

Healthcare analytics can be described as gathering and examining data from the healthcare sector for knowledge acquisition and decision-making. Healthcare data analytics can strengthen operations, enhance patient care, and cut costs at the macro- and micro-levels. Medical bills, clinical information, patient behavior, and prescriptions can be included on this list.

This knowledge is helpful, but it is also challenging. The data, whether from electronic health records (EHR) or from monitoring real-time vital signs, is derived from many different sources and must abide by government rules, making it a difficult and dangerous process. The knowledge, expertise, and abilities related to the ins and outs of the medical industry dominate the healthcare business. Along with the factors mentioned above, additional factors have an impact on the healthcare industry, such as decision-making accuracy and confidence in tasks and operations. This is because every action taken and decision made has the potential to change the course of events and have an impact on human health and life. Data analytics, a feature that spins the thread on which the healthcare industry functions, contributes to this accuracy. Healthcare is on its way to becoming another industry whose future is determined by data, due to the ongoing advancement of technology.

A data analytics and business intelligence (BI) solution can increase operational efficiency, lower costs, and streamline operations in any business by calculating and applying key performances indicator (KPIs) for decision-making and discovering opportunities. All participants, from payers and suppliers to patients and providers, can gain from the value of data [1].

Real world solutions (RWS) are based on thorough secondary research, combined with descriptive and predictive analytics. To provide our clients with effective solutions in the rapidly evolving fields of healthcare and life sciences, we have recognized the need for a unified strategy that integrates business expertise, scientific knowledge, and reporting services capabilities listed below.

- Analyses of patient journeys.
- Epidemiological research.
- Appraisal of the therapy area.
- Forecasting tools.
- Evaluation of business development and licensing.
- Knowledge of science.

Competitive intelligence-advanced analytics are used by businesses in the pharmaceutical and healthcare industries to make data-driven decisions in the increasingly cost-constrained global healthcare sector. Thanks to our unique method for determining an organization's analytical maturity and our flexible data management and solution suite, we provide various descriptive, predictive, and prescriptive analytics services for multinational clients [2].

1.1.1 INSIGHTS DERIVED FROM DATA THAT CAN ENHANCE PATIENT CARE

The speed of these changes is projected to quicken as the healthcare sector experiences numerous revolutionary shifts, such as adopting new EHR systems and procedures. Older methods of care are being replaced rapidly, and healthcare companies will need more efficient clinical data management to benefit from emerging, possibly game-changing trends in technology and analytics.

The following are some patterns that have emerged as a result of efforts to use big data in healthcare:
- Conversion from acute and episodic care models to value-based care.
- Increasing potential to profit from extensive databases of medical information.
- Robust analytics that might be used to address complex health issues.

For many individuals or organizations working in healthcare, simplifying and optimizing the process of collecting and arranging healthcare data can be an important initial action to take toward improving their overall operations or outcomes.

- Analytics using AI and machine learning in the healthcare sector.
- The best way to address the digital transformation of healthcare is with a well-informed strategy. The first step is to make use of the best tools available. Artificial intelligence (AI) and other automation tools are designed to amplify and enhance the work of experts. They aid in these experts' continual learning and speed up the pace of discovery and knowledge advancement.
- Given their capacity to reason, infer, and "understand" the interactions with their users, AI and machine-learning platforms are among those being

considered by healthcare providers top enterprises. These systems can take in enormous volumes of organized and unstructured data, and present the user with hypotheses to evaluate and a confidence rating for each insight and response.

1.1.2 TRANSFORMING DATA INTO VALUABLE IDEAS

To turn healthcare data into valuable insights, organizations will require specific information about their actual costs, the quality of the services they provide, and the actual relevance of those services [3].

By employing that knowledge, businesses can also adhere to encouraging trends, such as replacing quantity-based, fee-for-service models when delivering services with patient-centered, value-based care systems. Clinical IT is moving toward more centralized and integrated enterprise IT solutions.

Clinical equipment is becoming more networked, and data integration is becoming more automatic. Clinical and enterprise IT networks are connected through real-time location sensors and collaborative communication tools.

How real-time analytics enhance outcomes in the healthcare industry:

The state of global healthcare is unquestionably catastrophic due to high expenses, poor or inconsistent quality, and accessibility issues. The global population is getting older. There are currently more baby boomers among the elderly. In order to improve the quality of treatment at a significantly lower cost, the system is currently under significant financial pressure. The current trend of rising prices no longer linked to comfort is unsustainable [4–6].

1.1.3 PATIENTS' EXPECTATIONS ARE EVOLVING

The standard of care is becoming increasingly important as patients exercise their right to choose how and whom to work with for their healthcare. They want transparent procedures and data. Therefore, healthcare organizations will need to focus on figuring out how to communicate high-quality results in a way that is helpful to patients. Patient safety is highly valued by healthcare executives and organizations that advocate for patients. Detailed investigations of prescription errors, hospital-acquired infections, incorrect site procedures, and pressure sores (pressure sores, also known as pressure ulcers or bedsores, are injuries to the skin and underlying tissue that occur due to prolonged pressure or friction on a particular area of the body) will be required more than ever before.

Different ways in which medical interventions, technologies, and practices help to save the life of the patients:

1. Continuously fusing several data sources: Medical equipment shows visual representations of vital indicators via physiological streams, including electrocardiograms (ECG), heart rate, blood oxygen saturation (SpO_2), and respiratory rate. Initiatives to build electronic health records are expanding

the sources of medical data. Analytics that combine many data sources can be used to identify life-threatening illnesses, such as nosocomial infection, pneumothorax, intraventricular hemorrhage, and periventricular leukomalacia.

2. Highly individualized care: Identify warning indicators earlier to enhance patient outcomes and shorten hospital stays. Discovery of knowledge is automated or directed by clinicians to find novel connections between events in a data stream and medical conditions.

3. Proactive treatment: Create a patient profile for each patient using tailored data streams and receive ongoing insights.

1.2   APPLICATIONS OF COMPUTER VISION IN HEALTHCARE APPLICATION

Computer vision (CV) can be applied in healthcare for detecting different diseases.

1. Tumor detection: CV and deep learning applications have proven immensely helpful in the medical field, especially in detecting brain tumors accurately. Brain tumors spread quickly to other parts of the brain and spinal cord if left untreated, making early detection crucial to saving the patient's life. Medical professionals can use CV applications to make the detection process less time-consuming and tedious. In healthcare, CV techniques like Mask-R Convolutional Neural Networks (Mask R-CNN) can aid the detection of brain tumors, thereby reducing the possibility of human error to a considerable extent.

2. Medical imaging: CV has been used in various healthcare applications to assist medical professionals in making better decisions regarding the treatment of patients. Medical imaging or medical image analysis is one method that creates a visualization of particular organs and tissues to enable a more accurate diagnosis. Medical image analysis makes it easier for doctors and surgeons to glimpse the patient's internal organs to identify any issues or abnormalities. X-ray radiography, ultrasound, MRI, endoscopy, etc., are a few disciplines within medical imaging.

3. Cancer detection: Remarkably, deep learning CV models have achieved physician-level accuracy at diagnostic tasks such as identifying moles from melanomas. Skin cancer, for instance, can be challenging to detect in time, as the symptoms often resemble those of common skin ailments. As a remedy, scientists have used CV applications to differentiate between cancerous and non-cancerous skin lesions effectively. Research has also identified the numerous advantages of using CV and deep learning applications to diagnose breast cancer. Using a vast database of images of healthy and cancerous tissue, CV can help automate the identification process and reduce the chances of human error. With the rapid improvements in technology, healthcare CV systems can be used for diagnosing other types of cancer, including bone and lung cancer, eventually.

4. Medical training: CV is widely used not only for medical diagnosis but also for medical skill training. At present, surgeons depend on more than just the traditional manner of acquiring skills through actual practice in the operating room. Instead, simulation-based surgical platforms have emerged as an effective medium for training and assessing surgical skills. With surgical simulation, trainees can work on their surgical skills before entering the operating room. It allows them to gain detailed feedback and assessment of their performance, enabling them to better understand patient care and safety before actually operating on them. Computer vision can also assess the quality of the surgery by measuring the level of activity, detecting hectic movement, and analyzing time spent by people in specific areas.

5. The COVID-19 pandemic has posed a massive challenge to healthcare globally. With countries worldwide struggling with combating the disease, CV can significantly contribute to meeting this challenge. Due to rapid technological advancements, CV applications can aid in the diagnosis, control, treatment, and prevention of COVID-19. Digital chest X-ray radiography images, in combination with CV applications like COVID-Net, can easily detect the disease in patients. The prototype application, developed by DarwinAI in Canada, has shown results with around 92.4% accuracy in COVID diagnosis. CV is used to perform masked-face detection, which is widely used to enforce and monitor strategies preventing the spreading of pandemic diseases.

6. Health monitoring: CV and AI applications are increasingly used by medical professionals to monitor the health and fitness of their patients. These analyses allow doctors and surgeons to make better decisions in less time, even during emergencies. CV models can measure the amount of blood lost during surgeries to determine whether the patient has reached a critical stage. Triton, developed by AI-enabled platform effectively monitors and estimates the amount of blood lost during surgery. It helps surgeons to determine the amount of blood needed by the patient during or after the surgery.

7. Machine-assisted diagnosis: The advancement of CV in healthcare has led to more accurate diagnoses of ailments in recent years. The innovations in CV tools have proven to be better than human experts in recognizing patterns to spot diseases without error. These technologies are beneficial to help physicians identify minor changes in tumors to detect malignancy. By scanning medical images, such tools can aid the identification, prevention, and treatment of several diseases.

8. Timely detection of disease: For cancer, tumors, etc., the life and death of the patient depends on timely detection and treatment. Detecting the signs early gives the patient a higher chance of survival. CV applications are trained with vast amounts of data consisting of thousands of images that enable them to identify even the slightest difference with high accuracy. As a result, medical professionals can detect minimal changes that might have missed their eyes.

9. Home-based patient rehabilitation and monitoring: Many patients prefer to rehabilitate at home after an illness, compared to staying at a hospital. With CV applications, medical practitioners can provide patients with the necessary physical therapy and track their progress virtually. Such home training is not only more convenient but economical, too. In addition, CV technologies can also aid in the remote monitoring of patients or the elderly in a nonintrusive manner. A widely researched area is CV-based fall detection, where deep learning-based human fall-detection systems aim to reduce dependency and care costs in the elderly.

10. Lean Management in Healthcare: To correctly identify a disease, a medical professional must spend much time reviewing the reports and images to minimize the chances of error. However, implementing a CV tool or application can save a considerable amount of time while also getting highly accurate results. CV in healthcare helps hospitals to create maximum value for patients by reducing waste and waits. Queue detection, occupancy analysis, and people counting offer new tools to increase efficiency in healthcare. Unsurprisingly, many of those applications initially emerged in retail industries and are increasingly adopted by healthcare facilities to improve service quality and increase efficiency.

1.3   COMPUTER VISION CHALLENGES AND SOLUTIONS

Numerous sectors, such as healthcare, retail, and the automobile industry, have been transformed by CV technology. However, integrating CV into your company may be difficult and expensive, and poor planning might cause CV and AI projects to fail. Therefore, corporate managers must exercise caution before starting CV programs.

Four potential obstacles business managers may experience while adopting CV in their operations are discussed below, along with solutions for each.

1.3.1   INADEQUATE HARDWARE

Implementing CV technology involves using both software and hardware. A company must set up high-resolution cameras, sensors, and bots to guarantee the system's efficacy. This expensive technology may not be adequately fitted, resulting in blind spots and weak CV systems [7].

Some CV systems also need Internet of Things (IoT) enabled sensors; one study, for instance, shows how to use IoT-enabled flood monitoring sensors.

The following factors can be considered for effective CV hardware installation:

The cameras are high-definition and provide the required frames per second (FPS) rate. Cameras and sensors cover all surveillance areas.

The positioning covers all the objects of interest. For example, in a retail store, the camera should cover all the products on the shelf.

All the devices are appropriately configured to avoid blind spots.

1.3.2 POOR DATA QUALITY

1.3.2.1 Poor Quality

A CV system is built on top-notch tagged and annotated datasets. Since the effects of inaccuracy in CV systems can be detrimental, it is essential to have high-quality data annotation and labeling in sectors like healthcare, where CV technology is widely employed. For instance, many COVID-19 detection methods were unsuccessful because of insufficient data. Working with medical data annotation specialists can help mitigate this issue.

1.3.2.2 Lack of Training Data

There may be several difficulties in gathering adequate and pertinent data. These difficulties could result from a need for more training data for CV systems. For instance, collecting medical data can be difficult for data annotators. This is mainly because medical data is sensitive and private. Most medical photographs are either delicate or completely private, and hospitals and healthcare providers do not typically distribute them [8].

1.4 WEAK PLANNING FOR MODEL DEVELOPMENT

Another challenge can be weak planning when creating the machine learning (ML) model deployed for the CV system. During the planning stage, executives tend to set overly ambitious targets, which are difficult for the data science team to achieve.
Due to this, the business model:

- Does not meet business objectives.
- Demands unrealistic computing power.
- Becomes too costly.
- Delivers insufficient accuracy and performance.

Therefore, it is essential for business leaders to focus on the following:

- Creating a strong project plan by analyzing the business's technological maturity levels.
- Creating a clear scope of the project with set objectives.
- The ability to gather relevant data, purchase labeled datasets, or gather synthetic data.
- Considering the model training and deployment costs.
- Examining existing success stories similar to your business.

1.5 TIME SHORTAGE

During the planning phase of the CV project, business managers can tend to focus on something other than the model development stage. They fail to consider the extra time needed for:

- Setup, configuration, and calibration of the hardware, including cameras and sensors.
- Collecting, cleaning, and labeling data.
- Training and testing the model.

Image acquisition: The process of obtaining an image from sources. Hardware systems like cameras, encoders, and sensors can do this. It is, without a doubt, the essential phase in the machine vision(MV) workflow because a poor image would make the workflow shown in Figure 1.1 ineffective.

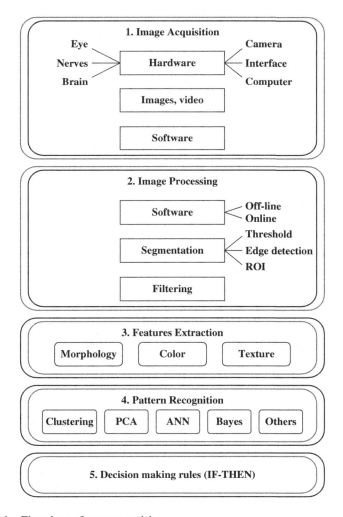

FIGURE 1.1 Flowchart of computer vision.

The steps in image acquisition are:

Step 1: Image acquisition: A sensor captures the image.
Step 2: Image enhancement.
Step 3: Image restoration.
Step 4: Color image processing.
Step 5: Wavelets.
Step 6: Compression.
Step 7: Morphological processing.
Step 8: Image segmentation.

The general aim of image acquisition is to transform an optical image (real world data) into an array of numerical data that can be later manipulated on a computer. Before any video or image processing can commence, an image must be captured by a camera and converted into a manageable entity.

1.6   INTRODUCTION TO IMAGE PROCESSING

Image processing is a technique for applying various procedures to an image to improve or extract relevant information from it. It is a type of signal processing where the input is an image, and the output can either be another image or features or characteristics related to that image. Image processing is one of the technologies that is currently expanding quickly. It is a primary research subject in engineering and computer science [9].

Image processing includes the following three steps:

• Importing the image via image acquisition tools.
• Analyzing and manipulating the image.
• Output in which the result can be altered or reported based on image analysis.

Analog and digital image processing are the two categories of image processing techniques. Hard copies like prints and images can use analog image processing. When applying these visual techniques, image analysts employ several interpretational fundamentals. Computer-based digital picture alteration is made possible using digital image processing tools. When employing the digital technique, all forms of data must go through three general phases: preprocessing, augmentation, presentation, and information extraction [10].

Feature extraction refers to the process of transforming raw data into numerical features that can be processed, while preserving the information in the original dataset. It yields better results than applying machine learning directly to the raw data.

Feature extraction can be accomplished manually or automatically.

Identification and description of the characteristics pertinent to a particular situation are necessary for manual feature extraction, as is implementing a method to extract those features. A solid grasp of the context or domain can often aid in deciding which characteristics might be helpful. Engineers and scientists have created feature extraction techniques for images, signals, and text through many years of research. The mean of a signal's window illustrates a specific feature.

Automated feature extraction uses specialized algorithms or deep networks to extract features automatically from signals or images without human intervention. This technique can be beneficial when you want to move quickly from raw data to developing machine learning algorithms. Wavelet scattering is an example of automated feature extraction [11].

The first layers of deep networks have primarily taken the position of feature extraction with the rise of deep learning, albeit primarily for picture data. Before developing powerful prediction models for signal and time-series applications, feature extraction is the first hurdle that demands a high level of knowledge.

Pattern recognition examines incoming data to look for patterns. While descriptive pattern recognition begins by classifying the identified patterns, explorative pattern recognition tries to identify data patterns in general.

Therefore, pattern recognition addresses both of these possibilities, and various pattern recognition techniques are used depending on the use case and type of data. As a result, pattern recognition is not a single technique but rather a large body of information and frequently unrelated skills.

The capacity to recognize patterns is frequently a need for intelligent systems. Words or texts, photos, or audio recordings can all be used as data inputs for pattern recognition. As a result, pattern recognition is more general than CV, which is more concerned with image identification.

Automatic and machine-based recognition, description, classification, and grouping of patterns are important problems in various engineering and scientific disciplines, including biology, psychology, medicine, marketing, CV, and AI [12].

Decision-making is the last step in the CV flow chart. Here, based on the image acquired and processed, and then after extraction of features and pattern recognition, the required decision is taken here at this stage.

1.7 THE FUTURE OF COMPUTER VISION

The applications of present-day CV seemed unachievable a few decades ago. Moreover, there is no end to the capabilities and future of CV technology.

1.7.1 A MORE COMPREHENSIVE RANGE OF FUNCTIONS

The capabilities of CV technology will expand as more research and development are done on it. In order to detect more images than it can present, the technology will be simpler to train. CV will also be combined with other technologies or branches of AI to produce more flexible applications. For instance, the environment can be understood by visually impaired people using a mix of picture-captioning software and natural language generation (NLG) [13].

1.7.2 LEARNING WITH LIMITED TRAINING DATA

The future of CV technologies lies in developing algorithms that require limited annotated training data compared to current models. To address this challenge, the industry has begun exploring a few potentially pioneering research themes:

Developmental learning: Machine systems that manipulate their surroundings to learn via a sequence of successes and failures when carrying out essential roles like navigating and grasping.

Lifelong learning: AI systems that capitalize on formerly learned visual concepts to attain new ones without direct supervision.

Reinforcement learning: Drawing inspiration from behavioral psychology, it concentrates on how robots can master how to take suitable actions.

CV technologies will soon join forces with robots in the physical world. Over the next decade, a key opportunity lies in developing robot systems that can smartly interact with human beings to help accomplish specific objectives.

1.7.3 LEARNING WITHOUT EXPLICIT SUPERVISION

As technology continues to advance, there is potential for significant improvements in computer vision and machine learning through the use of robots that can actively explore their surroundings. In the future, robots may be able to identify the objects in the images they observe, enabling them to autonomously move and follow the objects to gather multiple views without the need for manual labeling. This could enhance the efficiency and accuracy of data collection for machine learning algorithms, leading to more effective image recognition and classification capabilities [14].

1.7.4 ARTIFICIAL INTELLIGENCE AND MEDICAL ANALYTICS

The first step better health is finding smart tools. Solutions like cloud computing and healthcare analytics enable health data management, process automation, and data-supported decision-making in the healthcare industry.

The use of AI and machine learning technologies advances the situation. These systems can learn from various data types, including audio, video, and images. They have a large capacity for absorbing input, both structured and unstructured. They consider information and ideas from various sources before giving the user theories and predictive analytics to evaluate. Each insight and reaction on the platform are given a confidence level. Healthcare practitioners, researchers, and decision-makers can more rapidly see connections, correlations, and patterns relating to the issues they are attempting to solve and potential solutions once they have the results of those studies in hand [15].

1.7.5 UTILIZING ANALYTICS TO ENHANCE RESULTS

Healthcare firms increasingly use analytics regularly to find and apply new data insights. These strategies can be used to encourage clinical and operational improvements in order to overcome commercial issues.

Institutions in the healthcare industry are rapidly moving toward a predictive analytics-based paradigm. When data analysis and prediction capabilities take the place of information collection and report creation, analytics advances to the next logical level. In order to foresee future behavior, predictive analytics solutions combine historical data with modeling and forecasting. Additionally, they model potential outcomes. As a result, companies will be in a better position to identify fraud and provide more personalized approaches to consumer interaction and decision-making [16].

In the end, companies will want to be able to utilize predictive analytics' full potential to give decision-makers advanced tools for making decisions. The insights generated with agility, scalability, currency, breadth, and depth may impact results and enhance patient care, operational efficiency, and financial success.

1.8 IMPORTANCE OF HEALTHCARE DATA

In recent years, data gathering in healthcare settings has been streamlined. In addition to assisting in bettering daily operations and patient care, the data may now be used more effectively in predictive modeling. We can utilize both datasets to track trends and make forecasts rather than just focusing on historical or present data. We can now take preventative action and monitor the results.

The practice of providing healthcare on a fee-for-service basis is rapidly disappearing. In recent years, there has been a significant shift toward predictive and preventative interventions in terms of public health due to the increased need for patient-centric or value-based medical care. This is made feasible via data. Practitioners can identify individuals at high risk of acquiring chronic illnesses and assist in addressing a problem before it appears, instead of just treating the symptoms when they appear. Providing preventative care can help individuals avoid expensive hospitalizations and long-term health issues, which can result in lower costs for doctors, insurance providers, and patients alike. By investing in early detection and treatment of health conditions, healthcare providers can help prevent the development of more serious and costly health problems down the line. This approach can ultimately lead to better health outcomes and cost savings for all parties involved [17].

Data analytics can assist healthcare professionals in predicting the likelihood of infection, worsening, and readmission if hospitalization is necessary. The results of patient treatment can also be enhanced while expenditures are reduced.

1.9 BENEFITS OF HEALTHCARE DATA ANALYTICS

1.9.1 HEALTHCARE PREDICTIVE ANALYTICS

Predictive analytics have a significant and wide-ranging role in the healthcare industry, with applications far beyond enterprises. For instance, the US research group Optum Labs has compiled the electronic health records of over 30 million patients to create a database for predictive analytics tools that would improve healthcare. The main objective is to help clinicians make data-driven decisions faster and improve patient care. This is particularly true for patients with complicated medical histories and many illnesses. New technologies have the potential to identify individuals who may be at risk for certain health conditions, such as diabetes, and can provide them with information about the need for additional screenings or lifestyle changes, such as weight management. By using advanced algorithms and data analysis techniques, healthcare providers can generate personalized risk assessments and recommendations for patients, helping them to take proactive steps towards maintaining their health and preventing the onset of serious health issues.

1.9.2 TREATMENT OF PATIENTS AT HIGH RISK

Healthcare is typically expensive and complicated for people who need emergency care or services. Although it is not usually the case, expensive treatments can sometimes backfire by worsening patients' conditions. The digitization of medical information has made it easier to identify patient trends and histories. Predictive analytics can help identify patients more likely to go through crises due to their underlying medical issues. As a result, doctors have the opportunity to recommend preventative actions that lessen the need for emergency visits. Adopting a BI tool solution is critical in healthcare to keep high-risk patients secure, since correct data becomes necessary for assessing these patients and providing them with tailored care and treatment alternatives.

1.9.2.1 Patient Satisfaction

Concerns about the patient's level of engagement with the healthcare facility and satisfaction are justified. Wearables watch and other health monitoring devices enable doctors to give patients better preventative treatment while increasing their awareness of their role in their health. This knowledge accomplishes two goals simultaneously: Reducing hospitalization rates and fostering better communication between doctors and their patients. Using such devices can help prevent many health problems [18].

1.9.2.2 Industry Innovation

Data analysis tools have long-term benefits for the future growth of the healthcare industry, in addition to helping with immediate issues. Data analytics can be used to swiftly filter vast portions of data to find treatment options or remedies for various ailments, to provide accurate solutions based on previous data, and to permit personalized solutions to specific concerns for specific patients. For the growth of the healthcare industry, data analytics have a wide range of applications, including the prediction of epidemics, the treatment of diseases, the improvement of life quality, preventable care, prior diagnosis, and risk assessment.

1.9.2.3 Address Human Fallibility

Various avoidable health issues and insurance claims frequently result from human error, such as when a doctor prescribes the incorrect medication or dosage. As a result, individuals are exposed to more risk, and healthcare organizations are exposed to higher insurance costs and claim exposure [19]. Data analytics tools can be used to check patient data and medications delivered to validate data, alert users to any unexpected prescriptions or dosages, reduce human error, and prevent any occurrences of patient mortality or health issues.

1.9.2.4   Cost-cutting

Due to understaffing or overstaffing, treatment institutions, clinics, and hospitals frequently spend money on financial management. This irritation can be lessened by predictive analysis, which can help estimate the admission rate and guarantee that the proper staff is available to meet the patient's demands. As a result, patients can receive the care they require more promptly and effectively, with lower wait times. This also reduces the staffing requirements and the bed scarcity that hospitals frequently encounter due to insufficient financial management.

1.10   HEALTHCARE DATA ANALYTICS CHALLENGES

Because there need to be effective data governance rules in place, gathering data is one of the healthcare sector's most significant issues. The material must be brief, understandable, and formatted correctly to be used in various healthcare settings. Although it is common practice to store patient records in a centralized database for quick access, an issue arises when data must be shared with people outside the medical industry [20–22].

Data security is another growing challenge for healthcare providers because of the problem of frequent hacking and security violations that must be dealt with regularly. Caution and wisdom are needed while handling sensitive data, such as patient data, because any significant information leak might result in expensive losses [23].

1.10.1   THE GROWING ROLE OF THE HEALTH DATA ANALYST

The role of health data analysts is increasing data management, analysis, and interpretation abilities used by health data analysts to extract valuable insights from the data on healthcare being collected. The need for qualified health data analysts has expanded due to the use of big data in healthcare and the industry's growing drive for improvement [24–25]. Health data analysts do various tasks depending on their job and preferred industry. Regardless of the industry, a health data analyst must be able to work with, build, and evaluate health information technology (health IT) and other health information systems (HIS).

They can also be expected to:

- Collect or mine data
- Look at recent and historical data.
- Analyze uncooked data.
- Create forecasting models.
- Automatic reporting.

1.10.2   SKILLS NEEDED BY HEALTH DATA ANALYSTS

Analysts of health data provide their employers with solutions to difficulties. They should be proficient in the following skills in order to accomplish this:

- Structured Query Language (SQL).
- Data visualization using statistical programming in excel

As a health data analyst, one also needs to possess specific soft skills, such as:

- Rationality of thought.
- Sensible thinking.
- Strong attention to detail communication abilities.

1.10.3 WHERE WILL HEALTH DATA ANALYSTS BE EMPLOYED?

Many businesses and organizations use health data, and analysts are needed in these fields. These consist of the following:

- Public or commercial hospitals, as well as government healthcare divisions.
- Detection points.
- Healthcare insurance providers.
- Firms that consult on healthcare.
- Extensive medical facilities.
- Health IT companies.
- Various healthcare institutions.

Healthcare data analysts may work alone or on bigger teams, depending on the sector they choose to work in or the position they choose to fill.

1.11 SUMMARY

CV is a branch of artificial intelligence and deep learning where humans train computers to view and comprehend their surroundings. Even though humans and other animals deal with vision issues naturally from a very young age, helping robots detect and feel their surroundings through vision is still a primarily unresolved problem. Machine vision is inherently difficult because of the limitations of human eyesight and the ever-varying terrain of our dynamic world. Essentially, CV problems include teaching computers to comprehend digital images and visual information from the outside world. To do this, data from such sources may need to be gathered, processed, and analyzed to make decisions. The substantial formalization of complex problems into well-known, compelling problem statements is a defining feature of the development of machine vision.

Researchers worldwide can pinpoint problems and successfully address them because of the thematic division into categories, with precise definitions and appropriate vocabulary. Since the release of the ImageNet dataset in 2010, one of the subjects that has drawn the most attention is image categorization. Image categorization is a common task in computer vision, which can be performed by individuals with varying levels of expertise, from novices to specialists. With only a specified set of sample photos, the objective is to classify a given collection of images into a predefined set of categories. Image classification involves studying the entire image and labeling it, as opposed to more complex tasks like object identification and image segmentation, which need the localization of (or assignment of coordinates for) the traits they discover.

REFERENCES

1. Aghbashlo M, Hosseinpour S, Ghasemi-Varnamkhasti M (2014). Computer vision technology for real-time food quality assurance during the drying process. Trends Food Sci Technol 39:76–84.
2. Bonazzi C, Courtois F (2011). Impact of drying on the mechanical properties and crack formation in rice. In: Modern Drying Technology: Product Quality and Formulation (E Tsotsas and AS Mujumdar, eds.). Hoboken, NJ: Wiley-VCH, pp. 21–47.
3. Brosnan T, Sun DW (2004). Improving quality inspection of food products by computer vision—a review. J Food Eng 61:3–16.
4. Campeau C, Proctor JTA, Jackson CC, Rupasinghe HPV (2003). Rust-spotted north American ginseng roots: Phenolic, antioxidant, ginsenoside and mineral nutrient content. HortScience 38:179–182.
5. Campos-Mendiola R, Hernandez-Sanchez H, Chanona-Perez JJ, Alamilla-Beltran L, Jimenez-Aparicio A, Fito P, Gutierrez-Lopez GF (2007). Non-isotropic shrinkage and interfaces during convective drying of potato slabs within the frame of the systematic approach to food engineering systems (SAFES) methodology. J Food Eng 83:285–292.
6. Casleton DK, Shadle LJ, Ross AA (2010). Measuring the voidage of a CFB through image analysis. Powder Technol 203:12–22.
7. Velliangiri S, Karthikeyan P, Joseph IT, Kumar SA, (2019, December). Investigation of deep learning schemes in medical application. In 2019 International Conference on Computational Intelligence and Knowledge Economy (ICCIKE) (pp. 87–92). IEEE.
8. Chen Y, Martynenko A (2013). Computer vision for real-time measurements of shrinkage and color changes in blueberry convective drying. Dry Technol 31(10):1114–1123.
9. Chen YN, Sun DW, Cheng JH (2016). Recent advances for rapid identification of chemical information of muscle foods by hyperspectral imaging analysis. Food Eng Rev doi.10.1007/s12393-016-9139-110.
10. Courtois F, Faessel M, Bonazzi C (2010). Assessing breakage and cracks of parboiled rice kernels by image analysis techniques. Food Control 21(4):567–572.
11. Cubero S, Aleixos N, Molto E, Gomez-Sanchis J, Blasco J (2011). Advances in machine vision applications for automatic inspection and quality evaluation of fruits and vegetables. Food Bioprocess Tech 4(4):487–504.
12. Crank J (1975). The Mathematics of Diffusion, 2nd ed. New York: Oxford University Press.
13. Dalvand MJ, Mohtasebi SS, Rafiee S (2014). Optimization on drying conditions of a solar electrohydrodynamic drying system based on desirability concept. Food Sci Nutr 2(6):758–767.
14. Davidson VJ, Li X, Brown RB (2002). Fuzzy methods for ginseng drying control. In: The 9th International Conference on Information Processing and Management of Uncertainty in Knowledge-Based Systems, Paris, France, July 1–5 2002, 1–5, pp 913–918.
15. Davidson VJ, Martynenko AI, Parhar NK, Sidahmed M, Brown RB (2009) Forced-air drying of ginseng root: Pilot-scale control system for three-stage process. Dry Technol 27:451–458.
16. Demirhan E, Ozbek B (2009). Color change kinetics of microwave-dried basil. Dry Technol 27:156–166.
17. Du CJ, Sun DW (2004). Recent developments in the applications of image processing techniques for food quality evaluation. Trends Food Sci. Tech 15:230–249.
18. Du CJ, Sun DW (2006) Learning techniques used in computer vision for food quality evaluation: A review. J Food Eng 72:39–55.
19. Fan F, Ma Q, Ge J, Peng Q, Riley WW, Tang S (2013). Prediction of texture characteristics from extrusion food surface images using a computer vision system and artificial neural networks. J Food Eng 118:426–433.

20. Fernández L, Castillero C, Aguilera JM (2005). An application of image analysis to dehydration of apple discs. J Food Eng 67:185–193.
21. Karthikeyan P (2021). An efficient load balancing using seven stone game optimizations in cloud computing. Software: Pract Exp 51(6):1242–1258.
22. Karthikeyan P, Chandrasekaran M (2017). Dynamic programming inspired virtual machine instances allocation in cloud computing. J Comput Theor Nanosci 14(1):551–560.
23. Gomes GFS, Leta FR (2012). Applications of computer vision techniques in the agriculture and food industry: A review. Eur Food Res Technol 235(6):989–1000.
24. Gonzalez RC, Woods EE (2008). Digital Image Processing, 3rd ed. London: Pearson Education.
25. Goyache F, Bahamonde A, Alonso J, Lopez S, del Coz JJ, Quevedo JR, et al (2001). The usefulness of artificial intelligence techniques to assess subjective quality of products in the food industry. Trends Food Sci Tech 12(10):370–381.

2 Deep Learning Techniques for Foetal and Infant Data Processing in a Medical Context

Niha K
Amutha S
Surendiran B

CONTENTS

DOI: 10.1201/9781003345411-2

2.1 INTRODUCTION

Because of its compactness, lower cost, and noninvasive nature, ultrasound [1] imaging is a frequently utilized imaging modality for diagnosing, screening, and treating various disorders. During pregnancy, ultrasound imaging has proven to be the popular checkup modality. It is frequently used to measure a foetus's growth and development, monitor pregnancy, and assess clinical mistrust. Clinicians may find it challenging to analyze ultrasound images of the presence of features like speckle noise, acoustic shadows, and missing boundaries. Moreover, motion blurring is created due to the intricate interaction between ultrasound waves and the mother and foetus's biological tissues.

Deep learning (DL), particularly convolutional neural networks (CNNs), has played an increasingly important role in foetal ultrasound picture processing in recent decades. There is now a large body of literature on the subject. In the field of medical image analysis, several survey papers have been published in recent years, with a significant focus on the use of deep learning methods for ultrasound image analysis, particularly in the context of fetal sector analysis. Survey studies [2] dealing specifically with foetal ultrasound images include segmentation and classification algorithms; methods for foetal cardiology images; a summary of DL methods for identifying fetal abnormalities; and an analysis of research [3] from a clinical standpoint. Rationalized assessment of latest work in foetal ultrasound image analysis with DL can be a valuable and concise foundation for information. The first part of our study covers publicly accessible datasets from the subject area and widely accepted metrics for assessing the algorithm's performance.

The survey covers a wide range of topics, including foetal plane detection, anatomical structure analysis, and estimation of foetal biometry. These sections follow the steps used in clinical practice to evaluate the well-being of a fetus. Papers covering new tasks, ranging from less typical foetal evaluation applications to examination movement control, are collected in another area. Methods are outlined for each section, with pros and cons highlighted. Limitations and unresolved difficulties are explained further. The training and testing datasets, as well as the attained performance, are reported in summary tables. This study finishes with a conversation about DL regarding forthcoming prospects and difficulties in the field of fetal ultrasound analysis.

2.2 TRADITIONAL ULTRASOUND IMAGE ANALYSIS

High-frequency sound waves are employed with the technology of ultrasound imaging [4] to view the interior of the body. Since they are collected in real time, ultrasound scans can display blood flowing via blood vessels and internal organ movement. In contrast to X-ray imaging, you are not exposed to ionizing radiation during ultrasound imaging [5]. A transducer (probe) is positioned directly on the skin or inside a body orifice during an ultrasonic examination. A tiny layer of gel placed on the skin by the transducer acts as a conduit for the ultrasonic waves to enter the body. The ultrasonic image comprises waves reflected off the human body's

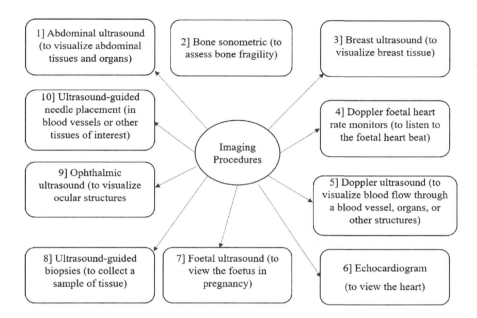

FIGURE 2.1 Ultrasound imaging procedures.

structures [6]. The power (amplitude) of the sound signal and the length of the wave's travel over the body offers the data necessary to build an image.

2.2.1 Uses

Ultrasound imaging [7] is a type of medical imaging that can help a doctor assess, diagnose, and treat medical problems. Standard procedures of ultrasound imaging are depicted in Figure 2.1.

2.2.2 Benefits/Risks

Since it has been used for more than 20 years, ultrasound imaging [8] has a solid reputation for safety. It does not provide the same risks as X-rays or other imaging technologies based on ionizing radiation, because it uses non-ionizing radiation.

Although ultrasound imaging is considered safe when carried out by appropriately trained medical professionals, ultrasonic energy can potentially have biological effects on the body. Ultrasound waves warm the tissues. This can sometimes result in small gas pockets in bodily fluids or tissues (cavitation). The long-term ramifications of these effects have yet to be discovered. Organizations [9] like the American Institute of Ultrasound in Medicine are concerned about the impact on the foetus. The ultrasonography used for obtaining foetal "keepsake" movies has also been discouraged. The photos or videos taken are acceptable for necessary examination.

Before undertaking any examination procedures, clinicians advise patients to talk with their doctor about the purpose of the examination, the medical data [10] which will be gathered, the risks involved, and how the results will be utilized to treat their condition or pregnancy. Ultrasound imaging is a suitable alternative for women of childbearing age who may otherwise be exposed to ionizing radiation from CT scans or other imaging modalities.

Ultrasound is a widely utilized medical imaging technique for examining the foetus [11] throughout pregnancy. The mother's and foetus's health are regularly evaluated and tracked. Ultrasound sessions [12] provide an opportunity for parents to connect with their unborn child by visualizing and hearing the fetus's heartbeat.

During a foetal ultrasound, a three-dimensional (3D) ultrasound [13] can be used to visualize a few facial features and other foetal organs like the toes and fingers. Four-dimensional (4D) [14] ultrasound is three-dimensional (3D) ultrasound in motion. While not many concerns are associated with ultrasonography, it is generally believed to be safe. However, risks can increase when exposure to ultrasound energy is too long or when unskilled individuals use the device.

The dangers of purchasing over-the-counter foetal heartbeat monitors should be clear to expectant mothers [15]. These gadgets should only be used by qualified healthcare professionals when medically necessary. Untrained users risk providing false information or exposing the foetus to dangerously high radiation levels. If medically necessary, healthcare practitioners [16] should seek exams that employ less or no ionizing radiation, such as an ultrasound or magnetic resonance imaging (MRI). According to laboratory tests, there are considerable implications. As a result, the FDA urges healthcare professionals who are using ultrasonography to consider measures to decrease exposure while maintaining diagnostic quality. In all imaging modalities, healthcare practitioners should adhere to the principles.

In accordance with the As Low As Reasonably Achievable (ALARA) guidelines, ultrasound procedures should prioritize the safety and efficacy of the devices used, with consideration given to the site and staff participation in accreditation and certification programs. These programs cover various aspects of device safety and performance, ensuring that ultrasound imaging is performed in a standardized and safe manner. In addition, manufacturers of electronic devices used in ultrasound imaging must comply with the regulatory requirements of Title 21 Code of Federal Regulations (1000 through 1005), which outlines the safety and performance standards for medical devices. These regulations aim to ensure that ultrasound devices are designed and manufactured in a way that minimizes potential risks to patients and healthcare providers [17]. Figure 2.2 depict the Federal regulation codes with description.

There are performance standards with non-federal radiation for diagnostic ultrasound. Ultrasound imaging devices conform to the regulations of medical devices [18] and are listed in Figure 2.3.

If adverse occurrences are reported right away, the FDA can identify and comprehend the risks connected to a product. Individuals and medical professionals are urged to voluntarily report any suspected issues with medical imaging devices to MedWatch, the FDA's Safety Information [19] and Adverse Event Reporting Program [20].

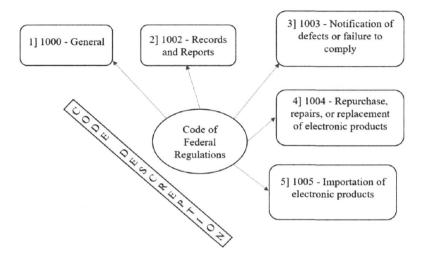

FIGURE 2.2   Federal regulation codes with descriptions.

Healthcare workers follow the policies established by their companies. Medical device producers, distributors, importers, and user institutions are subject to the Medical Device Reporting (MDR) Regulations of [21] CFR Part 803. In addition to adhering to the general guidelines for ultrasound imaging, the following details should be involved in reports, as illustrated in Figure 2.4.

The availability of the dataset for research in the field has been depicted in Table 2.1.

2.3   ARTIFICIAL INTELLIGENCE FOR FOETAL ULTRASOUND IMAGE ANALYSIS

Artificial intelligence (AI) is a vast field [22] that tries to use artificial ways to reproduce people's natural intelligence. AI approaches have recently become popular in the medical field. AI approaches are standalone approaches with indirect dependencies for medical imaging. With the advancement of new technologies, the concept of "joint decision-making" between AI and people can revolutionize healthcare AI subset approaches [23].

Computer-aided detection (CAD) diagnosis [24] has progressed to become an AI-assisted procedure in medical pictures, which includes the majority of medical imaging data, such as MRIs and CT scans. However, digitized medical images bring new data, opportunities, and challenges. As a result, AI approaches can overcome some of these issues by demonstrating outstanding sensitivity and accuracy in detecting imaging anomalies. These approaches could lead to diagnosis of illnesses by improving tissue-based detection and characterization. Furthermore, when using prenatal medical imaging [25], AI algorithms have demonstrated the ability to give promising results, such as tracking foetal development at each stage of pregnancy, forecasting the health of a placenta, and spotting potential issues.

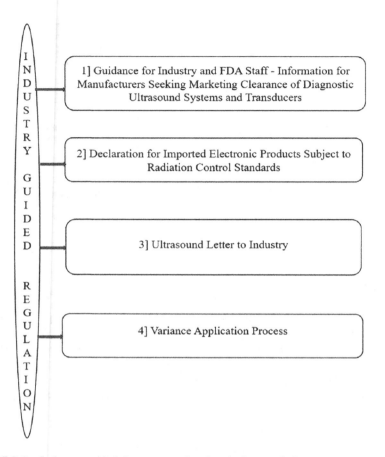

FIGURE 2.3   Industry-guided documents related to device regulations.

Congenital diaphragmatic hernias, foetal bowel obstruction, omphalocele, gastroschisis, pulmonary and sequestration, are examples of foetal diseases [26] that AI techniques may help detect. More research is needed to prevent and reduce adverse outcomes. Ultrasound imaging has gained popularity in medical research [27] and is now used during all three trimesters of pregnancy. Ultrasound is used to diagnose and track the foetus's growth and development. Ultrasound can also provide detailed anatomical information [28] about the foetus, high-resolution pictures, and improved diagnostic accuracy. When it comes to ultrasound, there are several advantages and a few drawbacks.

2.4   DEEP LEARNING FOR FOETAL ULTRASOUND IMAGE ANALYSIS

Deep CNNs have shown promise in medical imaging analysis of the foetus, including foetal brain segmentation. A fully convolutional network [29] has been utilized

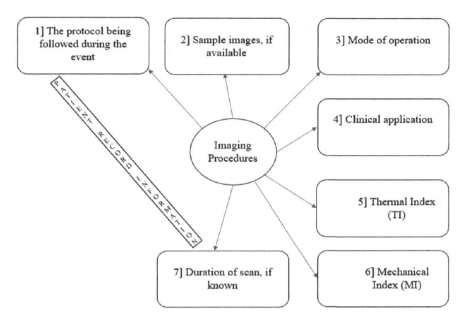

FIGURE 2.4 Information to list in patients' reports.

TABLE 2.1

The Data Set Available to Foster Research in Terms of Input Collection Method, Training Data Set Size, Testing Data Set Size, and Models Performance Metrics Are Listed Below

Name	Task(s) (Images/Patients)	Training Set Size (Images/Patients)	Testing Set Size	Performance Metrics
HC18 challenge dataset (2018)	Head-circumference estimation	999 2D/ *	335 2D/ *	DF [mm], ADF [mm], HD [mm], DSC [%]
A-AFMA ultrasound challenge dataset (2021)	1. MVP detection 2. Amniotic fluid and maternal bladder detection			mAP [0-1]
Burgosetal (2020) [20]	Standard plane detection of abdomen, brain, heart, femur, maternal cervix, and other diseases	7129 2D/ 896	5271 2D/ 896	top-1 error rate [%], top3errorrate [%], Acc [%]

to segment the foetal abdomen successfully. Several research studies have explored the use of convolutional neural networks (CNNs) for extracting fetal brain images. One study used a fully connected conditional random field to identify and extract fetal brain images [30]. Another study employed 2D U-Net and multistate U-Net models to perform whole-brain extraction from fetal images. These approaches utilize the power of deep learning algorithms to segment and extract the fetal brain, providing valuable insights into fetal development and potential health issues.

Skull segmentation [31] using a two-stage CNN has also been proposed, with the second stage containing angle incidence and shadow-casting maps. However, significant segmentation that quantifies multiple brain tissue classes is required for a fuller volumetric and morphologic assessment of the foetal brain. For multi-tissue foetal brain MRI segmentation, a 2D U-Net technique was developed. To improve the robustness of the segmentation, Khalili et al. employed data augmentation using simulated intensity inhomogeneity aberrations.

However, this strategy was only tested on a small group of people (n1/412). Payette et al. [32]. recently used the Fetal Tissue Annotation and Segmentation Dataset to test multiple two-dimensional (2D) segmentation algorithms. The combined incremental learning approach which comprised information from three independent 2D U-Net architectures (axial, coronal, and sagittal), performed the best of the deep learning models tested, indicating the advantage of using information from three planes. 3D U-Net uses anatomic information in three dimensions, avoiding segmentation failure caused by section discontinuity in 2D models. The 2D U-Net outperformed the 3D U-Net, which was attributed to the lower amount of training samples and the usage of non-overlapping patches in the 3D U-Net.

The adoption of uniform acquisition planes [33] increases the reproducibility of foetal biometry assessment and foetal evaluation, as per the International Society of Ultrasound in Obstetrics and Gynecology (ISUOG) standards. The foetal brain (FBSP), foetal abdomen (FASP), and foetal femur (FFESP) standard planes are commonly used to extrapolate biometric measures. The trans-thalamic standard plane (FTSP) and trans-ventricular standard plane (FVSP) are used in FBSP. Foetal heart (including four-chamber view [4CH]), left ventricular outflow tract (LVOT), right ventricular outflow tract (RVOT), three-vessel view (3VV), three-vessel trachea (3VT), foetal trans-cerebellum standard plane (FCSP), lumbosacral spine plane, and foetal facial standard plane (FFSP) are also required for foetal evaluation (FLVSP). The coronal, axial (FFASP), and sagittal planes are all included in the FFSP. Figure 2.5 shows visual examples of the most popular standard planes.

In practical practice, clinicians manually acquire a standard plane by moving the ultrasound probe around the mother's body until specified anatomical landmarks [34] become observable in the image. Due to differences in gestational weeks, kit manufacturers, and ultrasound-probe angles, clinical skill is essential to deal with the substantial intraclass changeability of ultrasound standard planes. Furthermore, the anatomical components that distinguish one plane from another may be shared by all planes.

The discussed DL techniques for standard-plane detection are summarized in Table 2.2. The initial techniques for detecting scan planes using deep learning [34] involved the use of two separate CNNs. The first CNN was designed to identify the

Fetal
Abdomen

Fetal Chamber
View

Fetal Femur

Maternal
Cervix

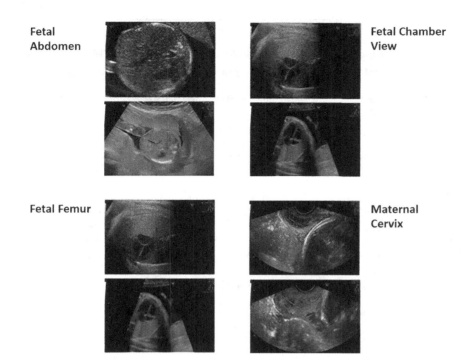

FIGURE 2.5 Anatomical landmarks of the baby using an ultrasound probe around the mother's body.

presence of the umbilical vein and stomach bubble within a localized area of the abdomen, while the second CNN was intended to detect both the stomach bubble and umbilical vein within the same localized area. Over 2,606 foetal abdominal pictures from 219 ultrasound movies, the system achieved a mean AUC of 0.99, with mean Acc as 0.98, Rec as 0.96, and Spec as 0.97.

Many techniques [35] treat the problem of detecting the foetal standard plane as a classification problem. In FFSP, a shallow classification CNN pretrained on ImageNet5 is fine-tuned to detect it. On 2,418 pictures, the method achieved an AUC of 0.99 and Acc, Prec, Rec, and F1 of 0.96, 0.96, 0.97, and 0.97, respectively.

A superficial classification CNN is also used to identify FBSPs automatically. To test the technique, the dataset contains 19,142 photos that have been supplemented and split into 5 folds. The CNN scores were 0.93, 0.93, 0.92, and 0.93 for Acc, Prec, Rec, and F1. Six planes [36] are classified by state-of-art CNNs (FTSP, FASP, FVSP, FCSP, and thorax standard planes). A dataset of 5,271 photos is used for testing (896 patients).

DenseNet-169 is the CNN with the best results, with Acc values of 6.20, 0.27, and 0.94, respectively. Additionally, 4CH, FASP, FBSP, and FFSPs are detected using a dense network [37]. The testing set consists of 5,678 ultrasound photos, with Prec, Rec, and F1 values of 0.98, 0.98, and 0.98, respectively. Also, the work suggests a DenseNet-based automatic classification of the 4CH, FBSP, FASP, FBSP, and

TABLE 2.2

The DL Techniques for Standard-Plane Detection, along with Training Data Set Size, Test Data Set Size, and Performance Metrics, Are Depicted Below

Paper (Year)	Plane	Training Set Size	Test Set Size	Performance Metrics
Qu et al. (2020)	FBSPs	15,314 2D (155 subjects)	3,828 2D (155 subjects)	Acc = 0.93, Prec = 0.93, Rec = 0.92, F1 = 0.93
Burgos-Artizzu et al. (2020)	Multiple	7,129 2D (896 subjects)	5,271 2D (896 subjects)	6.2% top-1 error, 0.27% top3 error, Acc = 0.94
Meng et al. (2020)	4CH, FASP, LVOT, RVOT, Lips, FFESP	12,000 2D	5,500 2D	F1 = 0.77, Rec = 0.77, Prec = 0.78
Montero et al. (2021)	FBSP	6,498 2D	2,249 2D	Acc = 0.81, AUC = 0.86, F1 = 0.80
Pu et al. (2021)	FASP, FTSP, FCSP, FLVSP	68,296 2D	16,740 2D	Acc = 0.85, Prec = 0.85, Rec = 0.85, F1 = 0.85
Cai et al. (2020)	FASP, FBSP, FFESP	Validation on 280 videos	Validation on 280 videos	Prec = 0.98, Rec = 0.85, F1 = 0.87
Lee et al. (2021)	Multiple	1,504 2D	752 2D	Prec = 0.75, Rec = 0.73, F1 = 0.74
Zhang et al. (2021)	FASP, FBSP, 4CH	2,460 2D	820 2D	mAP = 0.95, Acc = 0.95, Prec = 0.95, Rec = 0.93
Gao et al. (2020)	FTP, FCSP	34,586 2D from 441 videos (147 subjects)	60 videos (20 subjects)	mAP = 0.87
Yang et al. (2021)	Multiple	1,281 3D	354 3D	DF = 2.31 mm, θ = 10.36∘
Yang et al. (2021)	Uterine standard planes	539 3D (476 subjects)	144 3D (476 subjects)	DF = 1.82 mm, θ = 7.20∘
Tsai et al. (2021)	Middle sagittal plane	112 3D	28 3D	ED = 0.05

coronal FFSP as foetal standard planes. A placenta transfer dataset is used to train the network so that it may identify and understand potential relationships between the datasets, perhaps preventing overfitting. With Acc, Rec, Spec, and F1 values of 0.99, 0.96, 0.99, and 0.95, respectively, a test set of 4,455 images was employed.

A fully convolutional network (FCN), which seeks to segment the center of the foetal heart and classify the cardiac views in a single step, is used to address the identification of the four typical foetal heart planes (4CH, LVOT, 3VV, and not heart). 2,178 frames are used as the test set, yielding an error rate of 0.23 [38]. Generative adversarial network (GAN) enhance FBSP classification using ResNet. The approach is validated using a total of 2,249 pictures, yielding an AUC of 0.86 and Acc and F1 of 0.81 and 0.80, respectively.

Better feature orientation is used to extract discriminative and domain-invariant features through domains. The average in F1, Rec, and Prec for the performance in the target domain is 0.77, 0.77, and 0.78, respectively. A few academic articles include ultrasound video clips in the classification of common planes. Using a DL framework [39] is recommended to find FASP, FFASP, and 4CH. The framework uses a long short-term memory (LSTM) to analyze the temporal data encoded in the ultrasound videos. 331 movies from various subjects (a total of 13,247 pictures) are used for performance evaluation. The average Acc, Prec, Rec, and F1 for detecting planes is 0.87, 0.71, 0.64, and 0.64.

The study aimed to classify four different types of fetal scan planes, namely FASP, FTSP, FCSP, and FLVSP, using a proposed approach. To evaluate the effectiveness of the approach, a dataset of 224 films comprising 16,740 frames was used. The approach achieved mean accuracy, precision, recall, and F1 values of 0.85, 0.85, 0.85, and 0.85, respectively, indicating promising results in classifying fetal scan planes. Attention methods let CNN focus on discriminating regions in the ultrasound picture, which is used to enhance the performance of standard-plane detection and raise the interpretability [40] DL results. One of the earliest self-gated soft attention mechanisms for a CNN for identifying 13 ultrasound standard planes is added in this study. On 38,243 photos, the framework is tested, and mean values for Acc, Rec, Prec, and F1 are 0.98, 0.93, 0.93, and 0.93, respectively. A further option frequently used in the industry is a sonographer gaze attention mechanism.

For FASP detection, a multitasking architecture [41] relying on SonoNet is suggested. Both standard planar and sonographer visual saliency prediction are trained into the system. Eight films totaling 324 frames, each acquired from a different individual, are utilized to evaluate the method. The standard-plane detection performance is improved with sonographer visual saliency prediction, resulting in Prec, Rec, and F1 values of 0.96, 0.96, and 0.96, respectively. A similar strategy is put forth, which uses a temporal attention module to examine ultrasound temporal snippets further. The FASP, FBSP, and FFESP are considered planes. Following five-fold cross-validation, 280 movies ranging in length from three to seven seconds are divided, yielding mean Prec, Rec, and F1 values of 0.89, 0.85, and 0.87, respectively. It [38] uses various data augmentation techniques to enhance standard plane categorization. The method is validated using three-fold cross-validation on 1,129 standard plane frames (14 categories) and 1,127 background images. The results are Prec = 0.75, Rec = 0.73, and F1 = 0.74.

Anatomical-structure detection is a distinct form of standard plane detection that has been studied in the literature. A classification [42] CNN is used to identify images of the 4CH, and a multitask SSD is used to determine the essential anatomical components of the plane, as well as the ultrasound gain parameter and image zoom. Five-fold cross-validation on 7,032 photos is used to validate the algorithm. The mAP of the writers is 0.81. Certain anatomical structures in foetal head ultrasound images are assessed using a Faster R-CNN. A total of 1,153 photos were employed in the testing, with AP, Rec, and Prec values of 0.79, 0.85, and 0.87 to detect the structures of interest, respectively. The same strategy is used. Six important anatomical components are identified using a multitasking architecture that combines a Faster R-CNN with an additional classification branch, and it is determined whether the head is centered in the picture. Based on 320 test photos, an mAP of 0.93 is discovered.

A CNN inspired by Faster R-CNN is utilized in a multitasking framework [43] to predict the occurrence of particular anatomical structures in images of the heart, brain, and abdomen and to categorize these structures. The framework is tested on a total of 820 images, attaining mAP of 0.95 in recognizing structures and Acc, Prec, and Rec of 0.95, 0.95, and 0.93 in classifying them, respectively. A hybrid strategy is suggested, in which a CNN is taught to localize 13 anatomical structures with the help of image-level labels that provide poor supervision, negating the requirement for bounding-box annotation during training. Two hundred movies and 109,165 photos are used to test the methodology. Achieved are mean Prec, Rec F1, and IoU values of 0.77, 0.90, 0.80, and 0.62, respectively.

FVSP detection [44] is carried out by first localizing the foetal brain using a segmentation CNN. Then, using several other CNNs, CSP visibility, foetal brain symmetry, and midline orientation are evaluated. The framework is tested using a five-fold cross-validation on 19,838 photos, achieving an achievable FVSP detection in higher than 95% of instances. Scientists have studied self-supervised and semi-supervised standard-plane identification techniques. A semi-supervised pipeline for FTSP and FCSP detection from freehand foetal ultrasound video is given.

A CNN for feature extraction, a prototype learning module, and a semantic transfer module for automatically labeling buried video frames make up the framework [45]. The test set consists of 60 movies (20 participants), with an average mAP of 0.87. In this chapter, a self-supervised approach for detecting scan planes in foetal 2D ultrasound pictures is proposed. In particular, two small patches are randomly selected and switched on an image, and this technique is performed several times. A CNN is trained to return the changed image to its original state. The approach achieves Prec, Rec, and F1 of 0.89, 0.90, and 0.89, respectively, when tested on the same dataset as the one employed.

Utilizing a semi-supervised learning strategy [46], 13 typical planes are classified. A dataset of 22,757 photos is used as tagged data, with 100 images per class. The remaining pictures are handled as unlabeled information. Overall accuracy for a test set of 5,737 photos is very close to 0.70. Many researchers are attempting standard plane detection with DRL from 3D foetal ultrasound, mainly to emulate physicians' actions and investigate inter-plane dependency. The graft suggests a DRL identifies typical planes of the developing brain in ultrasound volumes. The DRL architecture includes a landmark-aware alignment module that uses a CNN to locate anatomical standards in the ultrasound volume. A plane-specific atlas is then recorded with the landmarks.

An Recurrent Neural Network (RNN) module ends the DRL agent's interaction process [47] in this way: On 100 ultrasound volumes, the effectiveness of the approach is validated, and an average theta of 9.36 and a mean DF of 3.03 mm are produced. The strategy is improved more by designing an adaptive RNN-based termination module that halts the agent search early. The approach was validated on a total of 100, 110, and 144 volumes of foetal brain, abdomen, and uterus, respectively, resulting in a mean DF of 2.31 mm and a mean of 10.36. A comparable method is used, in which a multi-agent DRL is used to simultaneously localize numerous uterine standard planes in 3D. A one-shot neural architecture search (NAS) module is included in the latter. A landmark-aware alignment approach is used to improve the robustness of the system in a challenging setting.

An RNN learns the spatial relationship between standard planes. One hundred forty-four volumes of foetal uterus and brain are used to test the procedure. The average and DF are 7.20 and 1.82 mm, respectively. 2D picture planes are sent to a CNN, which predicts the 3D transformation for registering each plane, and ultrasound volumes are processed. The method is examined using 3D ultrasound volumes of the foetal brain from 22 participants. A mean plane center [48] difference of 3.44 mm and a rotation angle of 11.05 degrees between the planes are found.

2.5 SURVEY STRATEGY

The following research questions, shown in Figure 2.6, served as the basis of a survey strategy.

A list of keywords for this survey was created [49] based on these questions, including "classification," "segmentation," "detection," "foetal," "ultrasound," "deep learning," and terminology relevant to foetal inspection and organs. SpringerLink, IEEEXplore, Scopus, ScienceDirect, and PubMed were the research databases used. A thorough review of the reference list was conducted for each of the resulting papers.

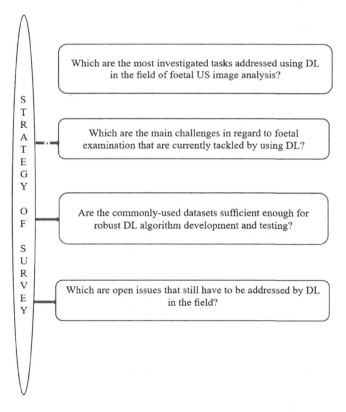

FIGURE 2.6 Questions related to the survey strategy on foetal examination using AI techniques.

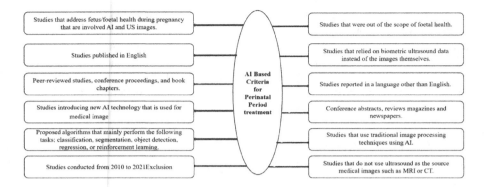

FIGURE 2.7 Criteria for inclusion and criteria outlined for computer vision AI treatments in the perinatal period.

We examined indexed journal and conference articles released in 2017 to enhance focus on recent and intriguing trends while avoiding overlay with earlier work.

Preferred Reporting Items for Systematic Reviews [50] and Meta-Analyses [51] (PRISMA) were used to report the findings. We based our early search terms on past research and consulted with a digital health expert to identify additional terms. This survey comprised studies on "foetus" or "foetal" published in the previous 11 years. In this investigation, we used numerous synonyms for computer vision AI treatments in the perinatal period. Criteria for inclusion and removal are outlined in Figure 2.7.

2.6 PUBLICLY AVAILABLE DATA SETS

It is not easy to collect and share high-quality [52] annotated foetal ultrasound datasets. Labeling huge datasets can take a long time, depending on the task (for example, pixel-level labeling for segmentation takes a lot longer than image-level labeling for classification). Concerns about data privacy and security are sometimes a deterrent to data sharing. International scientific organizations are attempting to collect and exchange publicly obtainable databases to boost algorithm development and reasonable comparison of algorithms to mitigate these concerns.

Three labeled datasets have been made available as part of the Grand Challenge [53] by the International Conference on Medical Image Computing and Computer-Assisted Intervention (MICCAI) and the IEEE International Symposium on Biomedical Imaging (ISBI). "Biometric Measurements from Foetal Ultrasound Images" was the first challenge held with ISBI 2012. The purpose was to autonomously segment the foetal abdomen, head, and femur so that conventional biometric data could be measured. Despite the dataset's importance, its size (270 photos) prevents researchers from developing generalizable DL algorithms.

The potential of DL for the biometry parameter [54] estimate was awakened in 2018 with the HC18 challenge dataset. With 999 and 335 2D ultrasound pictures for training and testing, the goal was to construct algorithms that automatically quantify foetal head circumference (HC). A skilled sonographer and a medical researcher annotated the images. The A-AFMA ultrasound challenge was held during ISBI 2021.

The objectives were to 1) detect amniotic fluid and the maternal bladder and 2) identify the proper landmarks for measuring the maximum vertical pocket (MVP) to determine the volume of amniotic fluid. A huge dataset of routinely obtained maternal foetal screening images [55] from the United States has been made accessible. It includes 7,129 2D training photos from 896 patients, divided into six categories: belly, brain, femur, thorax, maternal cervix, and others. An expert foetal clinician manually labeled the images. A test set was provided, which included 5,271 2D photos from 896 patients. The idea was to encourage more studies on foetal standard-plane detection.

2.7 HEART-BASED FOETAL ULTRASOUND ANALYSIS

The importance of foetal cardiac examination in detecting heart problems [56], such as congenital heart diseases (CHDs) and intrauterine growth constraint, is critical. Cardiac examination entails a cardiac function study and anatomical evaluation of the heart, including heart size and form. Heart and heart-structure detection are the focus of several DL techniques in the field. Five articles published in 2020–2021 that used machine learning approaches to analyze heart disease data are featured in this section. According to this study, the random forest method, which has an accuracy rate of over 88 percent, is the most successful strategy for producing correct results. The support vector machine technique [57], which achieved a result of above 85 percent in data analysis, was found to have the best accuracy in this investigation. Salhi et al. from Algeria did a study in 2021 that used three data analytics approaches [58] for analyzing data for persons with heart disease: nearest neighbors, neural networks, and support vector machines. According to their findings, the neural networks technique is the best for accuracy, with a 93 percent accuracy rate. Jindal et al. advise using machine learning approaches [59] for predicting and classifying people with heart disease.

Investigations on 1991 4CH reveal an mAP = 0.93. A localization algorithm for cardiac structures is suggested. Using a modified VGG-16, the existence of the heart in the 4CH is distinguished. Next, the occurrence of the mitral valve, foramen ovale, tricuspid valve, RV wall, and LV wall is temporally classified using a Faster R-CNN model combined with LSTM layers [60]. The movies of two out of 12 subjects, totaling 91 (39,556 frames), are used as the test set. The ACC value is 0.82. A recurrent CNN forecasts the foetal heart's presence, viewing plane, location, and orientation.

A total of 91 movies from 12 participants' prenatal cardiac screening tests have been annotated at the frame level. The accuracy of classification and localization is assessed using a 12-fold leave-one-out cross-validation over each individual. A custom-based metric [61] that considers classification and localization outcomes is suggested. The dataset includes 91 videos from 12 people, including two test videos. For incorrectly classifying views, the ACC is 0.83, and for inappropriately localizing structures, the ACC is 0.79.

Using an end-to-end two-stream CNN for temporal sequence analysis, spatial and temporal representations of the foetal heart are learned [62]. The objective is to supervise the acquisition of motion and appearance features weakly. Test data includes 15% of 412 ultrasound foetal films. A 0.90, 0.85, and 0.89 of ACC, Prec, and

Rec, respectively, are obtained for heart identification. Since the shape and structure of the heart are significant predictors of potential pathology, structure segmentation provides more information than detection. Encoder-decoder CNNs are frequently used in the literature for this purpose. To identify potential structural heart problems early, foetal cardiac standard planes are segmented using U-Net architecture. The test data consists of 106 pictures, some showing ventricular and atrial septal abnormalities.

IoU and Acc are calculated to be 0.94 and 0.96, respectively. The network comprises two stacked U-Nets and a dilated sub-network that is in charge of gathering global and local information [63]. The method is validated using five-fold cross-validation on 895 4CH views (895 healthy women). The average DSC, Acc, and AUC are 0.83, 0.93, and 0.99, respectively. Additionally, anatomical heart structures are segmented using a cascaded U-Net, which achieves DSC, HD, and Acc values of 0.85, 3.33, and 0.93 on 428 pictures, respectively. The first cardiac frame has labels to help the CNN be more precise. As the frames are successively segmented, the CNN is dynamically fine-tuned using shallow tuning to fix the most recent frame. The base points of the mitral valve were traced to identify the left ventricle and atrium regions. The testing set consists of 41 sequences, each with 40 frames.

The CNN achieves corresponding HD, AD, and DSC values of 1.260, 0.2, and 0.94. Several articles [64] accomplish instance segmentation of heart structures, one of which segments the four cardiac chambers using a network with three branches: the mask branch, the category branch, and a category-attention branch. This latter is utilized to improve segmentation accuracy and correct instance misclassification. After five-fold cross-validation, 638 pictures from 319 foetuses are used for testing. With an mAP of 0.45, the mean DSC for the four cardiac chambers, is 0.79, 0.76, 0.82, and 0.75, respectively.

The suggested method [65] is utilized to evaluate foetal cardiac problems (atrial septal, atrioventricular septal, and ventricular septal defects). As a test set, 10% of the 693 photos are used. Aorta DSC = 0.84, hole DSC = 0.68, LA DSC = 0.88, RA DSC = 0.89, and DSC = 0.87 are obtained. To look for potential heart anomalies, the foetal heart is segmented into its four conventional views—3VT, 4CH, LVOT, and RVOT—using Mask R-CNN. The testing collection consists of 116 ultrasound photos in total. This method gives the result as mAP = 0.96, IoU = 0.79, and DSC = 0.90.

In identifying cardiac structures in foetal standard views, such as the left ventricle and right atrium, an IoU of 0.72 is obtained. Another common task examined in the literature on foetal cardiac analysis is CHD diagnosis [66]. A detector for hypoplastic left heart syndrome (HLHS) has been created. The detector includes an aVGG16 to distinguish between HLHS patients and healthy participants and a SonoNet for detecting common planes (4CH, LVOT, RVOT). Data curation and split are manually carried out, starting with ultrasound pictures from 100 foetuses. The number of foetuses and frames in the test set is not stated. Prec, Rec, F1, and AUC are 0.85, 0.86, 0.86, and 0.93, respectively.

YOLOv2, which recognizes the ventricular septum, and U-Net, utilized to segment the cropped ventricular septum area, are used to analyze ventricular septal abnormalities. The output of U-Net is further improved with a calibrating module [67]. Two hundred eleven foetuses yield a total of 615 photos (421 videos). Validation is performed using three-fold cross-validation. IoU and DSC are obtained to be 0.55 and 0.68, respectively. YOLOv2 is also employed for the 4CH and 3VT video-based identification of cardiac anomalies. With the aid of three-fold cross-validation, the analysis is carried out on 2D frames taken from 34 videos. The detection of cardiac substructures is accomplished with an mAP of 0.70, and the assessment of heart structural anomalies is evaluated with a mean AUC of 0.83.

The classification of CHD patients and healthy participants is proposed using a one-class classification network. For data augmentation, a GAN [68] improvement is applied. The method is tested against state-of-the-art approaches using a balanced dataset of 400 test photos. The result is an Acc of 0.85. The first step is to identify the four-chamber views using a YOLOv3. Afterward, samples are classified as ED or ES using a mobile network. One hundred fifty-one movies altogether are utilized to assess the framework. Acc Rec, Spec, and F1 are 0.95, 0.93, 0.94, and 0.95, respectively.

The purpose is to identify various anatomical structures in various foetal cardiac images. The pipeline is meant to be a self-organized system to incorporate new data and is based on natural hierarchies in ultrasound videos. The test set consists of two of the 12 subjects' movies, a total of 91 (39,556 frames). The results highlight the benefits of incremental learning, which are presented with unique metrics.

2.8   DISCUSSION OF THE STUDY STRATEGY

This evaluation looked at many of the most recent DL algorithms [69] for analyzing foetal ultrasound images. The study of foetal images in the United States stretches back to the mid-nineteenth century, and there is now substantial and well-developed literature on the subject. The goal of our survey was to find answers to the subsequent query: What are the most researched tasks in the field of foetal ultrasound image analyses that use DL? Standard plane detection (19.9%) and biometry parameter estimation (21.9%) are two of the maximum explored tasks, according to our findings.

The foetal brain, abdomen, and heart standard planes have been recently studied in standard plane detection, since biometric measures and abnormality detection are mostly performed on these planes. The key to evaluating [70] the quality of each scanned plane is the identification of specific anatomical landmarks; hence, detection algorithms or attention processes are very useful in this sector, and the workflow is shown in Figure 2.8. In terms of biometry parameter estimates, the HC is a measurement that has received the most attention.

The most commonly used methods are segmentation (perhaps combined with head localization preprocessing), then fitting methods. Anatomical structure analysis was covered in 46.6% of the papers analyzed. Applications to the foetal heart and brain account for 30.3% and 24.2% of the publications, respectively. By initially identifying the three primary foetal cardiac planes—4CH, LVOT, and RVOT—detection rate [71] is frequently maximized. The analysis is frequently carried out using 2D ultrasound pictures or combining spatiotemporal data.

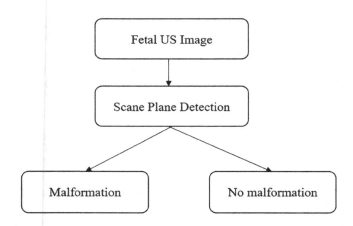

FIGURE 2.8   The overall workflow of a computer-aided tool to assist foetal clinicians.

Structure segmentation is primarily evaluated with architectures inspired by encoder-decoders, while the analysis is carried out with both 2D and, more recently, 3D structures. The examination of the foetal brain also consistently takes into account gestational age and brain development. The lungs, kidneys, and spine are the anatomical parts [72] that have gotten the least attention, and their techniques could be more varied. Current DL methods can analyze single-center datasets that were mostly collected and annotated by one or two ultrasound machines and are resilient for standard plane detection. The algorithms can deal with issues including a high degree of variation in foetuses' size and shape between different GAs and the existence of common anatomical components across various standard planes.

In the context of anatomical-organ competition, DL has established itself as a crucial technique to promote the segmentation of various anatomical entities, which can differ in size as well as shape over GA. That is not notable for anatomical structures contrasted with the background. Boundary incompleteness is a prevalent issue. Until this point, all supervised DL segmentation techniques [73] have outperformed other cutting-edge techniques in terms of performance and speed. Thus, dilated convolutions and residual learning are frequently used to boost the CNN field of view and, correspondingly, improve information loss.

For the past few decades, researchers have also worked on computer-aided diagnosis systems that fully utilize DL to aid physicians in their decision-making. DL currently obtains reliable results on HC estimate for biometry-parameter estimation, overcoming difficulties like the head's various positions in the image, the foetal head's varied sizes over the gestational trimesters, and the skull's partial visibility [74]. Only one publication in the field put forth a method resistant to a sizable shift in picture distribution between types of images.

When a small data sample is available, cross-validation [75] is strongly advised when we are dealing with multiclass issues. Cross-validation procedures test model capability using all available data, improving the reliability of algorithm performance. Which are the outstanding problems that DL must yet resolve in the field?

We next go over the main outstanding questions and potential future research areas in foetal ultrasound image analysis.

1. Multi-expert image annotation: A significant flaw at the moment is that different clinicians are unable to annotate images. While it is undeniably costly—both financially and in terms of time—to have many expert annotators, this multi-expert annotation is essential for the creation of reliable deep learning algorithms and the objective evaluation of algorithm performance. Additionally, the degree of picture complexity could be determined by evaluating the inter-clinician variability. When hand annotation is employed as the primary reference for the study, interpretation of results must be carefully assessed in light of these factors.

2. Performance evaluation: A crucial issue was the need for a systematic evaluation workflow. The irregular application of performance measures and testing datasets frequently makes it difficult to compare algorithms fairly. Since each job is validated by a distinct metric, a direct comparison of methodologies could not be made for all the sections being examined. In addition, the number of publicly accessible datasets is still insufficient despite the efforts of some scholars and international organizations, and they are frequently only made available for certain activities.

3. Comprehensive analysis: It is particularly difficult to create detailed computer models of the foetus, because as the child develops, its body constantly changes due to major structural and physiological changes between trimesters that affect inter- and intra-organ variability. Thus, one of the largest problems is to develop anatomically correct models that can capture the intricacy of foetus anatomy. Additionally, acquiring high-quality photos is the first stage in properly evaluating foetus well-being, because an end-to-end strategy capable of accurately classifying foetal planes and evaluating biometrics or foetal defects successively has yet to be fully utilized.

4. Semi-, weakly, and self-supervised learning: Researchers in related domains are suggesting semi-supervised, weakly supervised, or self-supervised ways to mitigate the problem of having tiny annotated datasets. Only 13 out of 145 studies in the foetal ultrasound domain examine the possibilities of such methods. These include investigating cross-device adaptation issues and classifying and detecting planes. The supervision of scan plane detection, representation of the fetal heart, multi-organ analysis, and shadow identification needs to be strengthened to improve their effectiveness. Scan plane detection, foetal position estimate, probe movement estimation, and foetal 3D reconstruction are all self-supervised approaches.

5. Model efficiency: The bulk of the surveyed studies do not mention the computational costs involved in the development and implementation of DL models. A high computational cost hampers the effective implementation of DL algorithms on single-board computers for on-the-edge computation. A high computational cost and a high CO_2 consumption go hand in hand. The performance in this area should be taken into consideration as an extra evaluation metric, because model efficiency is a crucial factor.

6. On-device DL in foetal ultrasound: The biggest tech companies in the world have embraced DL to offer cutting-edge goods and services in industries, including social networks, autonomous vehicles, and finance. Although DL applications for medical imaging produce excellent academic findings, commercial usage of these applications is still uncommon. SonoLyst is the first completely integrated AI solution in the world that enables the identification of the 20 views advised by the ISUOG mid-trimester practice recommendations for foetal sonography imaging over the DL algorithms used for commercial purposes.

7. Use of federated learning: The studies surveyed rely on datasets from a single center or datasets made accessible over global initiatives (e.g., Grand Challenge, Sec. II). It is still very difficult to obtain sufficiently expansive and varied databases of ultrasound photos of the foetus. Despite the advantages of such a paradigm, the studies under review have yet to use it.

8. Adherence to the ethics guidelines for trustworthy AI: Ethical standards and guidelines have become essential due to the quick growth of DL algorithms for foetal ultrasound image processing. The European Commission released a white paper in 2018 titled "Ethics Guidelines for Trustworthy AI" to highlight the significance of upholding and advancing ethical norms throughout all DL-related processes, from design to deployment. Still, efforts have yet to be made in this manner among the publications that have been evaluated. Researchers should complete this assessment list and report it as supplemental information.

2.9 CASE STUDY

2.9.1 HEART FOETAL ANALYSIS USING MACHINE LEARNING

The approaches used in this chapter, as well as the data collected and organized, are briefly discussed in this section. The machine that will be utilized to do this task has the following specifications in Figure 2.9.

The data reports [76] for 170 cases were collected and stored in an Excel file, with 13 features for each report. These characteristics display the numerical value. Machine learning techniques are primarily used in this study to forecast four diseases: (1) Arrhythmia, (2) cardiomyopathy, (3) CHDs, and (4) CAD. These strategies will be discussed briefly, and their significance will be clarified. The support vector machine creates a forecast for each entry based on the features that distinguish it. The output classes are associated with or dependent on a specific category. The hyperplane is sometimes produced by support vector machines (SVMs). However, the requisite qualities do not always result in a perfect separation spectrum. It results in a model that cannot be applied to other data. Table 2.3 shows the study, along with its values. Furthermore, SVM is a subset of learning algorithms designed to discriminate a binary qualitative variable. The model is to predict variables. Discrimination is considered a dichotomous variable.

Artificial neural networks [77] are a subset of artificial intelligence technologies that are employed in a wide range of applications today and have a bright future ahead of them. They are named after mathematical models that attempt to emulate

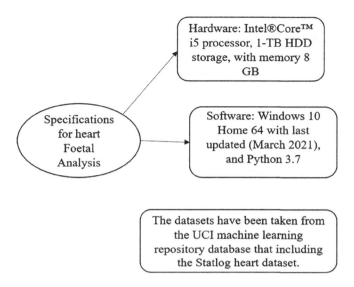

FIGURE 2.9 Specifications utilized to do heart foetal analysis using machine learning.

biological neurons. Furthermore, ANN is exemplified by its ability to tackle the most complicated issues in various fields, including prediction, optimization, pattern recognition, and others. It uses graphs and functions as a computational model. Because forward propagation is the initial round in dataset training, it was used in this study. Following that, all datasets are analyzed to ensure high prediction accuracy.

TABLE 2.3
A Medical Data Set on Heart Disease

Attribute	Limit
Age	Endless
Gender	Male = 0, Female= 1
Chest pain type	1 to 4
blood pressure	Endless
Serum cholesterol (mg / dl)	Endless
Satiety sugar level> 120 mg / dl	1 = True, 0 = False
Maximum heart rate	Endless
Electrocardiograph level at rest (0, 1, 2)	1 to 2
ST value at rest	Endless
Chest pain caused by exercise	1 = Yes, 0 = No
Number of major vessels (0–3)	Endless
The inclination of the ST segment in the peak exercise state	1 to 2
Damage Ratio: 3 = normal; 6 = permanent damage; 7 = reversible damage	3, 6, and 7

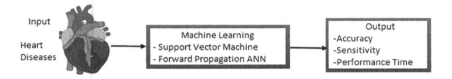

FIGURE 2.10 The mechanism of action starting with the process of inputting and processing data and ending with the display of the findings.

Figure 2.10 depicts overview of the proposed approach. The accuracy of the strategies [78] in forecasting each disease and the time it takes to implement each prediction procedure are depicted in Table 2.4. The ideal and acceptable implementation of each strategy in accurately diagnosing cardiac patients are depicted in Table 2.5.

The accuracy and sensitivity of each technique are calculated using the scientific formulas below, and the descriptions are given in Figure 2.11.

$$Accuracy = TP + \text{TN} / TP + FP + FN + TN \qquad (2.1)$$

$$Sensitivity = TP / TP + FN \qquad (2.2)$$

Two approaches are used in this chapter to predict people based on records. The data from 170 people is divided into 90 learning datasets and 80 testing datasets. To anticipate each disease, the accuracy and sensitivity of both approaches are evaluated. The best technique for achieving high-accuracy results, based on the experimental effects, forecasting, and assessing their performances, different techniques will be used in the future to forecast other cardiac ailments using the same data.

2.9.2 FOETAL ANALYSIS USING DEEP LEARNING

Our study's objective was to evaluate the new dataset against existing categorization methods [79]. To achieve this, we organized patients according to the date of their first visit, utilized all of the images from the first half of the patients to train the classifiers, and used the images from the second half of the patients to assess and report

---

TABLE 2.4

The Accuracy of the Strategies in Forecasting Each Disease, as Well as the Time It Takes to Implement Each Prediction Procedure

	SVM			FP-ANN		
Diseases	Accuracy	Sensitivity	P.T.	Accuracy	Sensitivity	P.T.
CHD	83.10%	81.20%	0.07912	72.70%	68.90%	0.065125
Cardiomyopathy	80.20%	77.10%	0.024425	85.60%	85.60%	0.07813
CAD	71.20%	68.10%	0.07813	69.60%	61.50%	0.012525
Arrhythmia	89.10%	89.10%	0.07912	85.80%	85.40%	0.024425

TABLE 2.5

The Ideal and Acceptable Implementation of Each Strategy in Accurately Diagnosing Cardiac Patients

Diseases	Optimal Execution	Acceptable Execution
CHD	SVM	FP-ANN
Cardiomyopathy	FP-ANN	SVM
CAD	SVM	FP-ANN
Arrhythmia	SVM	FP-ANN

the performance of the techniques. Consequently, a testing set of 5,271 images and a training set of 7,129 images were created (both with 896 patients).

Simple starting points: We initially contrasted two non-DL classifiers to determine how challenging it was? The multiclass boosting algorithm is used as a learning

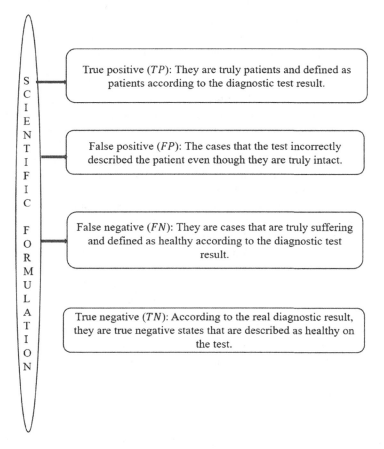

FIGURE 2.11 The descriptions of scientific formulas.

algorithm. The first processes the image pixels using principal component analysis, while the second uses a range of topologies, depths, full parameters, and processing speeds before being fully retrained (allowing network modifications) with our training data [80]. After that, the nets were applied to the test images using a frozen state, producing full probability scores for each class. If the loss on the validation set did not improve for five consecutive epochs, net training was terminated after a maximum of 15 epochs.

Learning rates were changed depending on the network, in accordance with the rules set forth by the network's original author. Weight decay was 0.9, and the batch size was 32. During training, data augmentation was employed to boost learning. Images were flipped at random, cropped from 0 to 20%, translated between 0–10 pixels, and rotated among [15: 15] degrees in each batch. ResNets and DenseNets were developed in TensorFlow [81]. MatConvNet was used for the rest, and pre-trained Imagenet models were retrieved directly. Excluding Inception, which utilizes a [299 299] image size, other nets used a [224 224] input picture size.

Because the images were grayscale, the initial convolutional layer of all nets was trained on Imagenet and customized to work on one-channel and three-channel filters. Finally, each network's fully connected layer [82] was modified to produce six classes. After convolution layers, batch normalization was applied to all nets. There was no more weight regularization implemented. The norm of all training images was used to transform input ultrasound images from an integer in the range [0, 255] to single float precision in the range [−1].

Discussions examined categorization in this work. For the first time, an expert physician gathered a big dataset of maternal-foetal ultrasound and meticulously classified it. A wide range of CNN was tested in a setting that simulated a genuine clinical scenario, and their performance was directly compared to that of research workers who conduct this activity on a daily basis using the same data. Because images were gathered by several operators using some ultrasound machines, the dataset represents a good portrayal of a true clinical scenario. The dataset is, as far as we know, the largest ultrasound dataset that has been made freely available to date to support research on automatic ultrasound recognition systems.

According to the results [83], it may be possible to identify maternal-foetal ultrasound using cutting-edge computer models created for broad visual identification from standard photos. Moreover, the computational model can classify photos 25 times faster and work 24 hours a day, seven days a week, implying that its employees could result in higher cost-effectiveness than utilizing technicians. However, the present technology has reached a saturation point, given the vast range of setups benchmarked and the narrow variation in performance. It may yet be beneficial to investigate fresh strategies specialized for usage on ultrasound photos in order to apply these hopeful results to other applications. The results of fine-grained classification [84] in the brain reveal a different picture: Computational models still need to catch up in activity, and more research in this area is required. Another issue to consider is the portability of computational models.

As per the results, a computational model trained solely from photos from the dataset accurately depicts a real-world clinical setting, and we expect that its public availability will spur more research. Here, 19 distinct CNNs (with various architectures and sizes) as well as two traditional machine learning methodologies are tested.

Both challenges directly compared outcomes against two separate research technicians. Also different components and training/testing processes looked at to improve and encourage more research.

As recognize that many different approaches might have been benchmarked. To summarize, computational models [85] have been shown for the first time to be mature enough for extensive use in real maternal-foetal clinical settings. In particular, there are still some issues around its application to fine-grained categorization, which is necessary for modern clinical diagnosis. This study, as well as maternal-foetal data from the United States, will serve as catalysts for future research.

2.10   CONCLUSION

In clinical patient care, the diagnostics of ultrasound imaging is very important. Because of its exceptional picture identification performance, DL, particularly the CNN-based model, has recently attracted much attention. If CNN lives up to its promise in radiology, it should help achieve perfection in diagnostics and improve patient care. AI-based diagnostic software [86] is important in academics and research as well as in the media. Diagnostic pictures are analyzed extensively in these systems.

In comparison to other radiological imaging technologies, ultrasound imaging, which is noninvasive, inexpensive, and non-ionizing, has restricted AI applications. As noted, the survey discovered one AI-based application for identifying foetal cardiac substructures [87] and indicating structural problems. A great deal of research based on these data has shown promise in diagnosing foetal illnesses or identifying certain foetal features. In the direction of the best of our facts, it is observed that no randomized controlled trial (RCT) or pilot research has been conducted in a medical center, nor has any AI-based application been adopted at hospitals.

Applying transfer learning [88] to a natively learned AI model on ultrasound images and fine-tuning the model on an advanced dataset gathered from a diverse medical center and/or diverse ultrasound equipment utilizing data augmentation to strengthen the generalization capability of DL models is another technique to overcome the tasks of a small dataset. Data augmentation settings should be carefully determined to recreate the changes in ultrasound images. In breast and chest imaging, advanced AI applications are already in use. Large amounts of medical images processed with solid reference principles are available in certain domains, allowing the AI structure to get trained. However, AI will have an impact on any medical application [89] that uses photos in the future. According to the survey, no standard guideline focusing on AI in diagnostic imaging is being followed or established to fulfill clinical context, evaluation, and criteria. As a start, critical characteristics to aid in the evaluation of AI research should be compiled. These suggestions are meant to improve the reliability [90] and use of AI research in diagnostic imaging.

2.11   FUTURE SCOPE

AI techniques have been used in various ways to expand fetal attention throughout this survey. Unfortunately, all components of these representations are not used as the product in medical institutions or hospitals because of the trustworthiness of

AI. Growing importance in medical student careers, the development of innovative training methodologies [91], there are currently no AI-based smartphone applications accessible to aid medical students, to the best of our knowledge. Aside from that, medical students may experience a variety of difficulties during foetal ultrasound scanning.

The conclusion is that AI approaches used ultrasound photos to monitor foetal health from various angles using GA. DL was the most commonly employed model in 107 investigations [92, 93] carried out using these models. Furthermore, we discovered that the foetus head received the most attention. Also recognized that the leading research institutes in this subject offer the dataset for each category is discussed in this survey, as well as their promising results.

Furthermore, it independently assessed to highlight the distinctive contribution of each study. In addition, for each category, DL's innovative and original works were highlighted. In conclusion, a future study approach has been suggested to address identified survey gap.

2.12 SUMMARY

Initially, this work was analyzed from the perspective of methodology and applications. We categorized the chapter into: traditional ultrasound analysis, AI for foetal ultrasound image analysis, DL for foetal ultrasound image analysis, survey strategy, publicly available datasets, heart-based foetal ultrasound analysis, discussion of foetal analysis, case studies of heart foetal analysis using machine learning, and foetal analysis using deep learning. The leading restrictions and disputes were discussed with tables for facilitating the evaluation of the different approaches. The use of publicly accessible datasets and performance measures to evaluate the effectiveness of algorithms was discussed. Finally, there was critical discussion of the current state of DL algorithms for foetal ultrasound image analysis, along with current challenges in the field which can be translated into the actual clinical practice.

REFERENCES

1. P. Zaffino, S. Moccia, E. De Momi, and M. F. Spadea, "A review on advances in intra-operative imaging for surgery and therapy: Imagining the operating room of the future," Annals of Biomedical Engineering, vol. 48, no. 8, pp. 1–21, 2020.
2. L. Meng, D. Zhao, Z. Yang, and B. Wang, "Automatic display of fetal brain planes and automatic measurements of fetal brain parameters by transabdominal three-dimensional ultrasound," Journal of Clinical Ultrasound, vol. 48, no. 2, pp. 82–88, 2020.
3. A. Ouahabi, and A. Taleb-Ahmed, "Deep learning for real-time semantic segmentation: Application in ultrasound imaging," Pattern Recognition Letters, vol. 144, pp. 27–34, 2021.
4. Y.-T. Shen, L. Chen, W.-W. Yue, and H.-X. Xu, "Artificial intelligence in ultrasound," European Journal of Radiology, p. 109717, 2021.
5. C. Song, T. Gao, H. Wang, S. Sudirman, W. Zhang, and H. Zhu, "The classification and segmentation of fetal anatomies ultrasound image: A survey," Journal of Medical Imaging and Health Informatics, vol. 11, no. 3, pp. 789–802, 2021.

6. P. Garcia-Canadilla, S. Sanchez-Martinez, F. Crispi, and B. Bijnens, "Machine learning in fetal cardiology: what to expect," Fetal Diagnosis and Therapy, vol. 47, no. 5, pp. 363–372, 2020.

7. Z. Chen, Z. Liu, M. Du, and Z. Wang, "Artificial intelligence in obstetric ultrasound: An update and future applications," Frontiers in Medicine, p. 1431, 2021.

8. X. P. Burgos-Artizzu, D. Coronado-Guti'errez, B. Valenzuela-Alcaraz, E. Bonet-Carne, E. Eixarch, F. Crispi, and E. Gratac'os, "Evaluation of deep convolutional neural networks for automatic classification of common maternal fetal ultrasound planes," Scientific Reports, vol. 10, no. 1, pp. 1–12, 2020.

9. Y. Cai, R. Droste, H. Sharma, P. Chatelain, L. Drukker, A. T. Papageorghiou, and J. A. Noble, "Spatio-temporal visual attention modeling of standard biometry plane-finding navigation," Medical Image Analysis, vol. 65, p. 101762, 2020.

10. L. H. Lee, Y. Gao, and J. A. Noble, "Principled ultrasound data augmentation for classification of standard planes," in International Conference on Information Processing in Medical Imaging. Springer, 2021, pp. 729–741.

11. R. Qu, G. Xu, C. Ding, W. Jia, and M. Sun, "Standard plane identification in fetal brain ultrasound scans using a differential convolutional neural network," IEEE Access, vol. 8, pp. 83821–83830, 2020.

12. Q. Meng, D. Rueckert, and B. Kainz, "Unsupervised cross-domain image classification by distance metric guided feature alignment," in Medical Ultrasound, and Preterm, Perinatal and Paediatric Image Analysis. Springer, 2020, pp. 146–157.

13. A. Montero, E. Bonet-Carne, and X. P. Burgos-Artizzu, "Generative adversarial networks to improve fetal brain fine-grained plane classification," Sensors, vol. 21, no. 23, p. 7975, 2021.

14. B. Pu, K. Li, S. Li, and N. Zhu, "Automatic Fetal Ultrasound Standard Plane Recognition Based on Deep Learning and IIoT," IEEE Transactions on Industrial Informatics, vol. 17, no. 11, pp. 7771–7780, Nov. 2021, https://doi.org/10.1109/TII.2021.3069470.

15. P.-Y. Tsai, C.-H. Hung, C.-Y. Chen, and Y.-N. Sun, "Automatic fetal middle sagittal plane detection in ultrasound using a generative adversarial network," Diagnostics, vol. 11, no. 1, p. 21, 2021.

16. X. Yang, H. Dou, R. Huang, W. Xue, Y. Huang, J. Qian, Y. Zhang, H. Luo, H. Guo, and T. Wang, et al., "Agent with warm start and adaptive dynamic termination for plane localization in 3D ultrasound," IEEE Transactions on Medical Imaging, 2021.

17. X. Yang, Y. Huang, R. Huang, H. Dou, R. Li, J. Qian, X. Huang, W. Shi, C. Chen, and Y. Zhang, et al., "Searching collaborative agents for multi-plane localization in 3D ultrasound," Medical Image Analysis, p. 102119, 2021.

18. Y. Gao, S. Beriwal, R. Craik, A. T. Papageorghiou, and J. A. Noble, "Label efficient localization of fetal brain biometry planes in ultrasound through metric learning," in Medical Ultrasound, and Preterm, Perinatal and Paediatric Image Analysis. Springer, 2020, pp. 126–135.

19. B. Zhang, H. Liu, H. Luo, and K. Li, "Automatic quality assessment for 2D fetal sonographic standard plane based on multitask learning," Medicine, vol. 100, no. 4, 2021.

20. S. Nurmaini, M. N. Rachmatullah, A. I. Sapitri, A. Darmawahyuni, A. Jovandy, F. Firdaus, B. Tutuko, and R. Passarella, "Accurate detection of septal defects with fetal ultrasonography images using deep learning-based multiclass instance segmentation," IEEE Access, vol. 8, pp. 196160–196174, 2020.

21. M. Rachmatullah, S. Nurmaini, A. Sapitri, A. Darmawahyuni, B. Tutuko, and F. Firdaus, "Convolutional neural network for semantic segmentation of fetal echocardiography based on four-chamber view," Bulletin of Electrical Engineering and Informatics, vol. 10, no. 4, pp. 1987–1996, 2021.

22. L. Xu, M. Liu, Z. Shen, H. Wang, X. Liu, X. Wang, S. Wang, T. Li, S. Yu, and M. Hou et al., "DW-net: A cascaded convolutional neural network for apical four-chamber view segmentation in fetal echocardiography," Computerized Medical Imaging and Graphics, vol. 80, p. 101690, 2020.

23. L. Xu, M. Liu, J. Zhang, and Y. He, "Convolutional-neural-networkbased approach for segmentation of apical four-chamber view from fetal echocardiography," IEEE Access, vol. 8, pp. 80437–80446, 2020.

24. S. An, H. Zhu, Y. Wang, F. Zhou, X. Zhou, X. Yang, Y. Zhang, X. Liu, Z. Jiao, and Y. He, "A category attention instance segmentation network for four cardiac chambers segmentation in fetal echocardiography," Computerized Medical Imaging and Graphics, vol. 93, p. 101983, 2021.

25. S. Nurmaini, M. N. Rachmatullah, A. I. Sapitri, A. Darmawahyuni, B. Tutuko, F. Firdaus, R. U. Partan, and N. Bernolian, "Deep learning-based computer-aided fetal echocardiography: Application to heart standard view segmentation for congenital heart defects detection," Sensors, vol. 21, no. 23, p. 8007, 2021.

26. J. Tan, A. Au, Q. Meng, S. FinesilverSmith, J. Simpson, D. Rueckert, R. Razavi, T. Day, D. Lloyd, and B. Kainz, "Automated detection of congenital heart disease in fetal ultrasound screening," in Medical Ultrasound, and Preterm, Perinatal and Paediatric Image Analysis. Springer, 2020, pp. 243–252.

27. A. Dozen, M. Komatsu, A. Sakai, R. Komatsu, K. Shozu, H. Machino, S. Yasutomi, T. Arakaki, K. Asada, and S. Kaneko et al., "Image segmentation of the ventricular septum in fetal cardiac ultrasound videos based on deep learning using time-series information," Biomolecules, vol. 10, no. 11, p. 1526, 2020.

28. M. Komatsu, A. Sakai, R. Komatsu, R. Matsuoka, S. Yasutomi, K. Shozu, A. Dozen, H. Machino, H. Hidaka, and T. Arakaki et al., "Detection of cardiac structural abnormalities in fetal ultrasound videos using deep learning," Applied Sciences, vol. 11, no. 1, p. 371, 2021.

29. R. Arnaout, L. Curran, Y. Zhao, J.C. Levine, E. Chinn, and A.J. MoonGrady, "An ensemble of neural networks provides expert-level prenatal detection of complex congenital heart disease," Nature Medicine, vol. 27, no. 5, pp. 882–891, 2021.

30. Z. Lv, "Rlds: An explainable residual learning diagnosis system for fetal congenital heart disease," Future Generation Computer Systems, vol. 128, pp. 205–218, 2022.

31. B. Pu, N. Zhu, K. Li, and S. Li, "Fetal cardiac cycle detection in multiresource echocardiograms using hybrid classification framework," Future Generation Computer Systems, vol. 115, pp. 825–836, 2021.

32. A. Patra, and J. A. Noble, "Hierarchical class incremental learning of anatomical structures in fetal echocardiography videos," IEEE Journal of Biomedical and Health Informatics, vol. 24, no. 4, pp. 1046–1058, 2020.

33. Y. Wu, K. Shen, Z. Chen, and J. Wu, "Automatic measurement of fetal cavum septum pellucidum from ultrasound images using deep attention network, "in 2020 IEEE International Conference on Image Processing, 2020, pp. 2511–2515.

34. V. Singh, P. Sridar, J. Kim, R. Nanan, N. Poornima, S. Priya, G. S. Reddy, S. Chandrasekaran, and R. Krishnakumar, "Semantic segmentation of cerebellum in 2D fetal ultrasound brain images using convolutional neural networks," IEEE Access, vol. 9, pp. 85864–85873, 2021.

35. L. Zhang, J. Zhang, Z. Li, and Y. Song, "A multiple-channel and atrous convolution network for ultrasound image segmentation," Medical Physics, vol. 47, no. 12, pp. 6270–6285, 2020.

36. M. K. Wyburd, M. Jenkinson, and A. I. L. Namburete, "Cortical plate segmentation using CNNs in 3D fetal ultrasound," in Medical Image Understanding and Analysis, B. W. Papiez, A. I. L. Namburete, M. Yaqub and J. A. Noble, Eds. Springer International Publishing, 2020, pp. 56–68.

37. L. Venturini, A. T. Papageorghiou, J. A. Noble, and A. I. Namburete, "Multitask CNN for structural semantic segmentation in 3D fetal brain ultrasound," Communications in Computer and Information Science, pp. 164–173, 2020.
38. L. S. Hesse, and A. I. L. Namburete, "Improving U-net segmentation with active contour based label correction," in Medical Image Understanding and Analysis, B. Papież, A. Namburete, M. Yaqub, and J. Noble, Eds. Communications in Computer and Information Science, vol 1248. Springer, 2020, https://doi.org/10.1007/978-3-030-52791-4_6
39. X. Yang, X. Wang, Y. Wang, H. Dou, S. Li, H. Wen, Y. Lin, P.-A. Heng, and D. Ni, "Hybrid attention for automatic segmentation of whole fetal head in prenatal ultrasound volumes," Computer Methods and Programs in Biomedicine, vol. 194, p. 105519, 2020.
40. F. Moser, R. Huang, A. T. Papageorghiou, B. W. Papie z, and A. I. Namburete, "Automated fetal brain extraction from clinical ultrasound volumes using 3D convolutional neural networks," Communications in Computer and Information Science, pp. 151–163, 2020.
41. H. N. Xie, N. Wang, M. He, L. H. Zhang, H. M. Cai, J. B. Xian, M. F. Lin, J. Zheng, and Y. Z. Yang, "Using deep-learning algorithms to classify fetal brain ultrasound images as normal or abnormal," Ultrasound in Obstetrics and Gynecology, vol. 56, no. 4, pp. 579–587, 2020.
42. L. H. Lee, E. Bradburn, A. T. Papageorghiou, and J. A. Noble, "Calibrated bayesian neural networks to estimate gestational age and its uncertainty on fetal brain ultrasound images," in Medical Ultrasound, and Preterm, Perinatal and Paediatric Image Analysis. Springer, 2020, pp. 13–22.
43. M. K. Wyburd, L. S. Hesse, M. Aliasi, M. Jenkinson, A. T. Papageorghiou, M. C. Haak, and A. I. Namburete, "Assessment of regional cortical development through fissure based gestational age estimation in 3D fetal ultrasound," in Uncertainty for Safe Utilization of Machine Learning in Medical Imaging, and Perinatal Imaging, Placental and Preterm Image Analysis. Springer, 2021, pp. 242–252.
44. Z. Hu, R. Hu, R. Yan, C. Mayer, R. N. Rohling, and R. Singla, "Automatic placenta abnormality detection using convolutional neural networks on ultrasound texture," in Uncertainty for Safe Utilization of Machine Learning in Medical Imaging, and Perinatal Imaging, Placental and Preterm Image Analysis. Springer, 2021, pp. 147–156.
45. V. A. Zimmer, A. Gomez, E. Skelton, N. Ghavami, R. Wright, L. Li, J. Matthew, J. V. Hajnal, and J. A. Schnabel, "A multitask approach using positional information for ultrasound placenta segmentation," in Medical Ultrasound, and Preterm, Perinatal and Paediatric Image Analysis. Springer, 2020, pp. 264–273.
46. H. C. Cho, S. Sun, C. M. Hyun, J.-Y. Kwon, B. Kim, Y. Park, and J. K. Seo, "Automated ultrasound assessment of amniotic fluid index using deep learning," Medical Image Analysis, vol. 69, p. 101951, 2021.
47. S. Sun, J.-Y. Kwon, Y. Park, H. C. Cho, C. M. Hyun, and J. K. Seo, "Complementary network for accurate amniotic fluid segmentation from ultrasound images," IEEE Access, vol. 9, pp. 108223–108235, 2021.
48. T.-H. Xia, M. Tan, J.-H. Li, J.-J. Wang, Q.-Q. Wu, and D.-X. Kong, "Establish a normal fetal lung gestational age grading model and explore the potential value of deep learning algorithms in fetal lung maturity evaluation," Chinese Medical Journal, vol. 134, no. 15, p. 1828, 2021.
49. P. Chen, Y. Chen, Y. Deng, Y. Wang, P. He, X. Lv, and J. Yu, "A preliminary study to quantitatively evaluate the development of maturation degree for fetal lung based on transfer learning deep model from ultrasound images," International Journal of Computer Assisted Radiology and Surgery, vol. 15, no. 8, pp. 1407–1415, 2020.

50. N. Weerasinghe, N. H. Lovell, A. W. Welsh, and G. Stevenson, "Multiparametric fusion of 3D power Doppler ultrasound for fetal kidney segmentation using fully convolutional neural networks," IEEE Journal of Biomedical and Health Informatics, vol. 25, no. 6, pp. 2050–2057, 2020.

51. A. Franz, A. Schmidt-Richberg, E. Orasanu, and C. Lorenz, "Deep learning-based spine centerline extraction in fetal ultrasound," in Bildverarbeitung Für Die Medizin 2021. Springer, 2021, pp. 263–268.

52. L. Chen, Y. Tian, and Y. Deng, "Neural network algorithm-based three dimensional ultrasound evaluation in the diagnosis of fetal spina bifida," Scientific Programming, vol. 2021, 2021.

53. E. L. Skeika, M. R. Da Luz, B. J. T. Fernandes, H. V. Siqueira, and M. L. S. C. De Andrade, "Convolutional neural network to detect and measure fetal skull circumference in ultrasound imaging," IEEEAccess, vol. 8, pp. 191519–191529, 2020.

54. J. C. Senra, C. T. Yoshizaki, G. F. Doro, R. Ruano, M. A. B. C. Gibelli, A. S. Rodrigues, V. H. K. Koch, V. L. J. Krebs, M. Zugaib, and R. P. V. Francisco et al., "Kidney impairment in fetal growth restriction: Three-dimensional evaluation of volume and vascularization," Prenatal Diagnosis, vol. 40, no. 11, pp. 1408–1417, 2020.

55. Q. Chen, Y. Liu, Y. Hu, A. Self, A. Papageorghiou, and J. A. Noble, "Cross-device cross-anatomy adaptation network for ultrasound video analysis," in Medical Ultrasound, and Preterm, Perinatal and Paediatric Image Analysis. Springer, 2020, pp. 42–51.

56. M. Alsharid, R. El-Bouri, H. Sharma, L. Drukker, A. T. Papageorghiou, and J. A. Noble, "A curriculum learning based approach to captioning ultrasound images," in Medical Ultrasound, and Preterm, Perinatal and Paediatric Image Analysis. Springer, 2020, pp. 75–84.

57. J. Perez-Gonzalez, N. H. Montiel, and V. M. Bañuelos, "Deep learning spatial compounding from multiple fetal head ultrasound acquisitions," in Medical Ultrasound, and Preterm, Perinatal and Paediatric Image Analysis. Springer, 2020, pp. 305–314.

58. Salhi, Dhai Eddine & Tari, Abdelkamel & Kechadi, Tahar. (2021). Using Machine Learning for Heart Disease Prediction. 10.1007/978-3-030-69418-0_7.

59. Jindal, Harshit & Agrawal, Sarthak & Khera, Rishabh & Jain, Rachna & Nagrath, Preeti. (2021). "Heart disease prediction using machine learning algorithms", IOP Conference Series: Materials Science and Engineering. 1022. 012072. 10.1088/1757-899X/1022/1/012072.

60. Wang, Lu & Guo, Dong & Wang, Guotai & Zhang, Shaoting. (2020). Annotation-Efficient Learning for Medical Image Segmentation Based on Noisy Pseudo Labels and Adversarial Learning. IEEE Transactions on Medical Imaging. 1-1. 10.1109/TMI.2020.3047807.

61. Y. Zeng, P.-H. Tsui, W. Wu, Z. Zhou, and S. Wu, "Fetal ultrasound image segmentation for automatic head circumference biometry using deeply supervised attention-gated v-net," Journal of Digital Imaging, vol. 34, no. 1, pp. 134–148, 2021.

62. D. Qiao, and F. Zulkernine, "Dilated squeeze-and-excitation U-Net for fetal ultrasound image segmentation," in 2020 IEEE Conference on Computational Intelligence in Bioinformatics and Computational Biology. IEEE, 2020, pp. 1–7.

63. M. G. Oghli, S. Moradi, N. Sirjani, R. Gerami, P. Ghaderi, A. Shabanzadeh, H. Arabi, I. Shiri, and H. Zaidi, "Automatic measurement of fetal head biometry from ultrasound images using deep neural networks," in 2020 IEEE Nuclear Science Symposium and Medical Imaging Conference. IEEE, pp. 1–3.

64. P. Bhalla, R. K. Sunkaria, A. Kamboj, and A. K. Bedi, "Automatic fetus head segmentation in ultrasound images by attention based encoder decoder network," in 2021 12th International Conference on Computing Communication and Networking Technologies. IEEE, 2021, pp. 1–7.

65. P. Li, H. Zhao, P. Liu, and F. Cao, "Automated measurement network for accurate segmentation and parameter modification in fetal head ultrasound images," Medical & Biological Engineering & Computing, vol. 58, no. 11, pp. 2879–2892, 2020.

66. M. C. Fiorentino, S. Moccia, M. Capparuccini, S. Giamberini, and E. Frontoni, "A regression framework to head-circumference delineation from ultrasound fetal images," Computer Methods and Programs in Biomedicine, vol. 198, p. 105771, 2021.

67. S. Moccia, M. C. Fiorentino, and E. Frontoni, "maskr2CNN: A distance-field regression version of mask-RCNN for fetal-head delineation in ultrasound images," International Journal of Computer Assisted Radiology and Surgery, vol. 16, no. 10, pp. 1–8, 2021.

68. Y. Meng, M. Wei, D. Gao, Y. Zhao, X. Yang, X. Huang, and Y. Zheng, "CNN-GCN aggregation enabled boundary regression for biomedical image segmentation," in International Conference on Medical Image Computing and Computer-Assisted Intervention. Springer, 2020, pp. 352–362.

69. J. Zhang, C. Petitjean, P. Lopez, and S. Ainouz, "Direct estimation of fetal head circumference from ultrasound images based on regression CNN," in Medical Imaging With Deep Learning. PMLR, 2020, pp. 914–922.

70. J. Zhang, C. Petitjean, F. Yger, and S. Ainouz, "Explainability for regression CNN in fetal head circumference estimation from ultrasound images," in Interpretable and Annotation-Efficient Learning for Medical Image Computing. Springer, 2020, pp. 73–82.

71. F. A. Hermawati, H. Tjandrasa, and N. Suciati, "Phase-based thresholding schemes for segmentation of fetal thigh cross-sectional region in ultrasound images," Journal of King Saud University-Computer and Information Sciences, 2021.

72. T. Singh, S. R. Kudavelly, and K. V. Suryanarayana, "Deep learning based fetal face detection and visualization in prenatal ultrasound," in 2021 IEEE 18th International Symposium on Biomedical Imaging. IEEE, 2021, pp. 1760–1763.

73. Y. Zhou, H. Chen, Y. Li, Q. Liu, X. Xu, S. Wang, P. T. Yap, and D. Shen, Multi-task learning for segmentation and classification of tumors in 3D automated breast ultrasound images," Medical Image Analysis, vol. 70, no. 101, pp. 918, 2021.

74. C. Chen, X. Yang, R. Huang, W. Shi, S. Liu, M. Lin, Y. Huang, Y. Yang, Y. Zhang, and H. Luo et al., "Region proposal network with graph prior and IoU-balance loss for landmark detection in 3D ultrasound," in 2020, IEEE 17th International Symposium on Biomedical Imaging. IEEE, 2020, pp. 1–5.

75. M. G. Oghli, A. Shabanzadeh, S. Moradi, N. Sirjani, R. Gerami, P. Ghaderi, M. S. Taheri, I. Shiri, H. Arabi, and H. Zaidi, "Automatic fetal biometry prediction using a novel deep convolutional network architecture," Physica Medica, vol. 88, pp. 127–137, 2021.

76. J. C. Prieto, H. Shah, A. J. Rosenbaum, X. Jiang, P. Musonda, J. T. Price, E. M. Stringer, B. Vwalika, D. M. Stamilio, and J. S. Stringer, "An automated framework for image classification and segmentation of fetal ultrasound images for gestational age estimation," in Medical Imaging 2021: Image Processing, vol. 11596. International Society for Optics and Photonics, 2021, p. 115961N.

77. K. Rasheed, F. Junejo, A. Malik, and M. Saqib, "Automated fetal head classification and segmentation using ultrasound video," IEEE Access, vol. 1, no. 1, pp. 99, 2021.

78. S. Bohlender, I. Oksuz, and A. Mukhopadhyay, "A survey on shape constraint deep learning for medical image segmentation," IEEE Reviews in Biomedical Engineering, 2021.

79. A. Casella, S. Moccia, D. Paladini, E. Frontoni, E. De Momi, and L. S. Mattos, "A shape-constraint adversarial framework with instance normalized spatio-temporal features for inter-fetal membrane segmentation," Medical Image Analysis, vol. 70, p. 102008, 2021.

80. S. Cengiz, and M. Yaqub, "Automatic fetal gestational age estimation from first trimester scans," in International Workshop on Advances in Simplifying Medical Ultrasound. Springer, 2021, pp. 220–227.

81. X. Chen, M. He, T. Dan, N. Wang, M. Lin, L. Zhang, J. Xian, H. Cai, and H. Xie, "Automatic measurements of fetal lateral ventricles in 2d ultrasound images using deep learning," Frontiers in Neurology, vol.11, p. 526, 2020.

82. F. Zhu, M. Liu, F. Wang, D. Qiu, R. Li, and C. Dai, "Automatic measurement of fetal femur length in ultrasound images: A comparison of random forest regression model and SegNet," Mathematical Biosciences and Engineering, vol. 18, no. 6, pp. 7790–7805, 2021.

83. J. Chen, Y. Zhang, J. Wang, X. Zhou, Y. He, and T. Zhang, "Ellipsenet: Anchor-free ellipse detection for automatic cardiac biometrics in fetal echocardiography," in International Conference on Medical Image Computing and Computer-Assisted Intervention. Springer, 2021, pp. 218–227.

84. S. Płotka, T. Włodarczyk, A. Klasa, M. Lipa, A. Sitek, and T. Trzciński, "Fetalnet: Multitask deep learning framework for fetal ultrasound biometric measurements," in International Conference on Neural Information Processing. Springer, 2021, pp. 257–265.

85. Y. Gao, L. Lee, R. Droste, R. Craik, S. Beriwal, A. Papageorghiou, and A. Noble, "A dual adversarial calibration framework for automatic fetal brain biometry," in IEEE/CVF International Conference on Computer Vision, 2021, pp. 3246–3254.

86. T. Włodarczyk, S. Płotka, P. Rokita, N. Sochacki-Wójcicka, J. W'ojcicki, M. Lipa, and T. Trzciński, "Spontaneous preterm birth prediction using convolutional neural networks," in Medical Ultrasound, and Preterm, Perinatal and Paediatric Image Analysis. Springer, 2020, pp. 274–283.

87. L. Zhang, T. Portenier, C. Paulus, and O. Goksel, "Deep image translation for enhancing simulated ultrasound images," in Medical Ultrasound, and Preterm, Perinatal and Paediatric Image Analysis. Springer, 2020, pp. 85–94.

88. Q. Meng, J. Matthew, V. A. Zimmer, A. Gomez, D. F. Lloyd, D. Rueckert, and B. Kainz, "Mutual information-based disentangled neural networks for classifying unseen categories in different domains: Application to fetal ultrasound imaging," IEEE Transactions on Medical Imaging, vol. 40, no. 2, pp. 722–734, 2020.

89. T. Liu, Q. Meng, A. Vlontzos, J. Tan, D. Rueckert, and B. Kainz, "Ultrasound video summarization using deep reinforcement learning," in International Conference on Medical Image Computing and Computer Assisted Intervention. Springer, 2020, pp. 483–492.

90. H. Sharma, L. Drukker, P. Chatelain, R. Droste, A. T. Papageorghiou, and J. A. Noble, "Knowledge representation and learning of operator clinical workflow from full-length routine fetal ultrasound scan videos," Medical Image Analysis, vol. 69, p. 101973, 2021.

91. C. Zhao, R. Droste, L. Drukker, A. T. Papageorghiou, and J. A. Noble, "Visual-assisted probe movement guidance for obstetric ultrasound scanning using landmark retrieval," in International Conference on Medical Image Computing and Computer-Assisted Intervention. Springer, 2021, pp. 670–679.

92. M. Luo, X. Yang, X. Huang, Y. Huang, Y. Zou, X. Hu, N. Ravikumar, A. F. Frangi, and D. Ni, "Self context and shape prior for sensorless freehand 3D ultrasound reconstruction," in International Conference on Medical Image Computing and Computer-Assisted Intervention. Springer, 2021, pp. 201–210.

93. H. Santhiya, and P. Karthikeyan. "Survey on auction-based scheduling in grid and cloud environment." International Journal of Computer Applications, vol. 62, no. 8, pp. 6–9, 2013.

3 Analysis of Detecting Brain Tumors Using Deep Learning Algorithms

Geetha A
Shaheen H
Rajagopal R

CONTENTS

3.1 INTRODUCTION

The medical imaging process is an influential technology used to diagnose diseases and monitor their treatment. One of the leading causes of cancer-related mortality is the brain tumor, which is the uncontrolled growth of malignant cells

DOI: 10.1201/9781003345411-3 **51**

in or around the brain. One of the most active research areas in medical image analysis, accurate brain tumor categorization, is crucial for extracting therapeutically relevant data from magnetic resonance imaging (MRI). Several efficient brain tumor detection frameworks are identified to gaining valuable insights essential for diagnosis and disease treatment planning. Variation in dimension, form, and structure of the brain tumor is one of the significant issues in tumor classification [1].

3.2   DEEP LEARNING IN MEDICAL IMAGES

In the past several years, academics have focused much attention on deep learning, the newest and most popular trend in machine learning. Deep learning has been widely employed in multiple applications as a powerful machine learning method for handling complicated problems requiring exceptionally high accuracy and sensitivity, particularly in the medical industry. Brain tumors are among the most prevalent and severe malignant tumor disorders, and if they are identified at a more advanced stage, they might result in a shorter life and poorer prognosis. Accordingly, grading a brain tumor is a crucial step to take after finding the tumor in order to develop a successful treatment strategy [2].

Artificial intelligence neural networks are composed of many layers. Each layer acts as an image filter. There are numerous algorithms in deep learning, each with advantages and drawbacks. These algorithms cover almost all aspects of image processing, focusing mainly on classification and segmentation [3].

Image classification and its use in deep learning have advanced significantly. On the one hand, academic circles have worked hard to create several effective Convolutional Neural Network (CNN) models, which have even outperformed human recognition capabilities in terms of accuracy. On the other hand, one of the most popular applications of deep learning is using the CNN model in medical picture processing. Recently, there has been significant advancement in the retinal brain image obtained from picture classification and its application to deep learning [4].

Investigating linked brain tumor diseases is the primary method for categorizing and detecting benign and malignant tumor images using deep learning algorithms. The deep CNN is used to update iterations with the advancement of deep learning techniques, going from the initial shallow CNN model to the deep CNN model or other combination models, including migration learning, data augmentation, and other new approaches and techniques [5]. It mainly functions in transfer learning for tumor illness detection. Large-scale medical annotation sets are tough to come by, and transfer learning is an excellent way to deal with the problem of tiny data to identify the possible relationships between accuracy and the significant model type.

Figure 3.1 describes a deep belief network (DBN) type of deep neural network used in machine learning. It comprises numerous layers of latent variables, or "hidden units," with connections between the layers but not between the units within each layer.

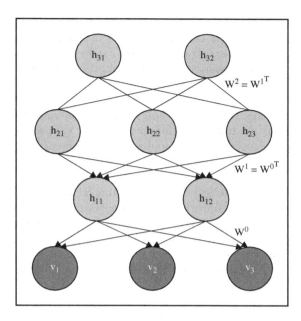

FIGURE 3.1   Deep belief networks.

3.3   LITERATURE REVIEW

A comprehensive analysis of the approaches found in the literature is described. It provides the nitty gritty for the methodological innovations provided in the following chapters. It contains demonstrative and quantitative studies employed in traditional brain tumor classification approaches. Indeed, innumerable transformative medical imaging systems and techniques have been employed to visualize, investigate, and understand brain structures to generate quicker and more accurate insights for making clinically valuable decisions in prognosis and treatment planning. This survey considers several MRI scan-based brain tumor detection models to provide a comprehensive overview and the theoretical background of previous works under five major topics of brain tumor detection: Preprocessing, segmentation, feature extraction, classification, and optimization. Then, we review the state-of-the-art brain tumor classification techniques in each phase to justify their use in this content.

Researchers developed a classification approach, using the Naive Bayes (NB) classifier, to accurately identify areas in the brain tumor that contain active cancerous cells. The study focused on brain cancer cell identification [6]. The proposed work integrates the Naive Bayes (NB) classification algorithm with denoising, pixel subtraction, morphological operations, maximum entropy threshold, and statistical feature extraction techniques to classify brain tissues. When the method is evaluated on 50 MRI scans, it achieves 100% prediction efficiency on normal brain scans and 81.25% efficiency on tumor scans with a classification accuracy of 94%.

Arunkumar et al. [7] developed an enhanced automatic brain tumor classification model by applying the concepts of ANN (artificial neural network). The authors employed MRI scans to detect brain cancers without human intervention. The proposed model consists of three important improvement techniques. Initially, this model utilizes a K-means clustering (KMC) algorithm to identify the exact area of cancer. Then ANN is used to select the optimal pixel by utilizing the knowledge gathered from the training process. Consequently, the segmentation process extracts the texture feature of the cancer region. To detect a tumor, this model utilizes greyscale features to diagnose and classify brain cancers. This model achieves 94.07% classification accuracy with 90.09% sensitivity and 96.78% specificity.

In general, CNN is an efficient machine learning technique to realize improved results in brain tumor identification. Badza and Barjaktarovic [8] proposed an effective CNN design for classifying three types of tumors. The proposed model is simpler than other pretrained models. The enactment of the model is assessed using extensive experimental analysis. This model achieves 96.56% of classification accuracy. With enhanced generalization ability and reduced time complexity, this model is used as an efficient decision support system for helping radiologists in clinical diagnostics. Havaei et al. [9] also developed a fully automated brain cancer classification model using CNN architecture that concurrently utilizes both local and global contextual features. This model processes each slice separately and correlates each pixel to different MRI modalities. The proposed CNN model achieves maximum accuracy of 88%, with 89% specificity and 87% sensitivity

Gumaei et al. [10] proposed a cohesive feature extraction model with a Regularized Extreme Learning Machine (RELM) algorithm to identify and classify brain cancers. The approach uses a min-max normalization strategy to improve the contrast of brain areas and edges in an MRI scan. Subsequently, the features of brain tumors are extracted using a hybrid feature extraction approach. In the end, RELM is applied to classify brain tumors. In order to assess and relate the performance of the developed model, a set of experiments is carried out in an open-source database with MRI scans collected from several patients. This model provides improved results with 91.51% to 94.233% classification accuracy related to existing state-of-the-art approaches.

3.4   BRAIN TUMOR ISSUES AND CHALLENGES

Anything that increases your likelihood of developing a disease, like a brain or spinal cord tumor, is a risk factor. There are various risk factors for various cancers. You can alter some risk factors, such as smoking. Others are unchangeable, such as your age or family history. However, having one or more risk factors does not guarantee that a person will get the disease; many people develop brain or spinal cord tumors without having any recognized risk factors. Though they may share some characteristics, the many different forms of tumors that can begin in the brain or spinal cord may not share the same risk factors. Most brain tumors have no identified risk factors [8].

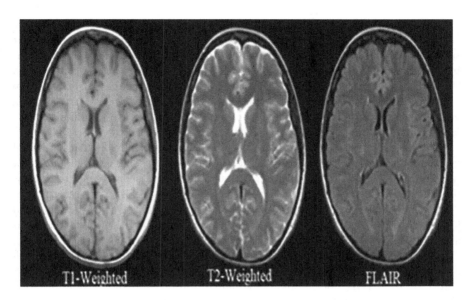

FIGURE 3.2   T1-weighted MRI, T2-weighted MRI and FLAIR.

To divide the sample into four groups, add the perpendicular diameters of all enhancing lesions and compare them to the smallest quantifiable tumor volumes. A magnetic resonance imaging (MRI) sequence called fluid-attenuated inversion recovery (FLAIR) sets the inversion recovery to null fluids. The categorization only considers T1-weighted sequences with contrast administration and, to a lesser extent, features on T2 and FLAIR images shown in Figure 3.2.

Figure 3.2 shows that stimulation by the separate relaxation processes of T1, or magnetization in the same direction as the static magnetic field, and T2, or magnetization in the opposite direction to the static magnetic field, each tissue (T) returns to its equilibrium state.

3.5   ANALYSIS STUDY

3.5.1   AN IMPROVED EXTREME LEARNING MACHINE WITH PROBABILISTIC SCALING

An extreme learning machine (ELM) is an effective learning algorithm used in single-hidden layer feedforward neural networks (SLFN), which arbitrarily chooses input parameters and bias constants of hidden neurons without executing the learning process [11]. The output weights are systematically calculated through the norm least-square solution and Moore-Penrose inverse of a conventional system, therefore enabling a significant reduction in time consumption.

This model employs the logistic (sigmoid activation) function and analysis the impact of the number of neurons in the hidden layer by varying the proportions of the number of features in the training and testing data. This simple learning process is analogous to conventional gradient descent algorithms concerning classification rate and root mean square error (RMSE) for brain cancer identification problems. Occasionally, this ELM may become unsuitable because of the arbitrary parameter selection and bias parameter among the hidden and input layers. Hence, the probabilistic scaling technique is used with ELM. The Extreme Learning Machine with Probabilistic Scaling (ELMPS) is a machine learning algorithm that differs from other neural networks in that it does not require random initialization in the hidden and input layers. Instead, it uses a probabilistic scaling approach to initialize these layers, which can lead to more accurate and efficient training.

The most significant problem with these learning algorithms is that the initial point assignment step lacks a proper technique and must be done manually or randomly [12]. Even though ELM-based techniques have been created to address this issue, their use is inappropriate, because these algorithms only offer a marginally better answer and require much time to implement. Due to the input and hidden layers' arbitrary parameter selection and bias, this ELM could occasionally be unsuitable. As a result, ELM employs the probabilistic scaling technique. This ELMPS disregards the idea of random parameter initialization in the hidden and input layers. The proposed classification model uses an enhanced ELMPS algorithm to identify normal and pathological brain tissues shown in Figure 3.3. The classification model includes a number of phases to identify brain tumors, including preprocessing, feature extraction, feature selection, similarity measure, and classification methods. A Wiener filter is used initially to remove extraneous pixels from the input image. A feature extraction method is used based on the images' contour, boundary, and density to extract pertinent characteristics from the MRI scan. Then principal component analysis (PCA) is used to reduce the number of features. The photos are classified using an enhanced ELMPS algorithm at the end.

3.5.1.1 Image Denoising

Many variables, including camera sounds, patient movement during image acquisition, and human and technical errors, influence the accuracy of brain MRI scans. Brain tumor identification requires simultaneously classifying numerous sites (pixels or objects) in a scan. In order to eliminate or reduce the noise present in the input scans, this work integrates maximum a posteriori (MAP) with a Markov random field (MRF).

3.5.1.2 Feature Extraction

Feature extraction methods greatly influence the classification accuracy of tumor detection algorithms. By establishing feature vectors, it pulls pertinent data about the condition of the brain from MRI scans. Then this feature vector is used to retrieve information and identify tumors. In this study, the structures, edges, and intensity of brain tumors are taken into consideration. The feature related to the tumor's structure provides information regarding the size and type of tumors. Boundaries are a characteristic that sheds light on the tumors' margins. The intensity-related features in the brain MRI scan show how each pixel's brightness varies. These three features are combined to form a feature vector. Based on that, features are extracted.

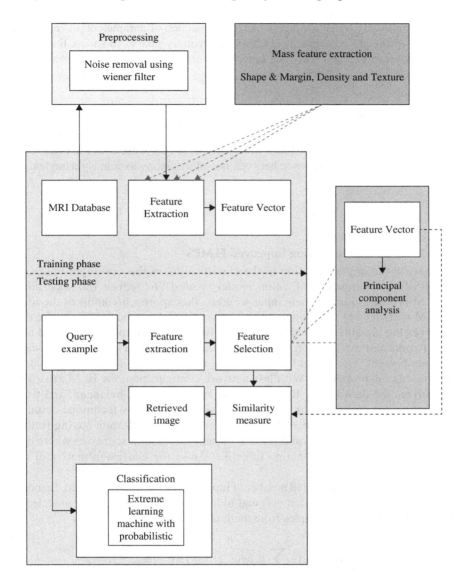

FIGURE 3.3   Architecture of a proposed ELMPS-based classification framework.

3.5.1.3   Feature Selection

In order to reduce the dimensionality of the data set without erasing the original brain image data, feature selection (reduction) is crucial in tumor detection models. As a result, the system's classification accuracy is maintained while the dimension of the feature space is reduced. Selecting a small group of the most important traits from the initial collection of features is the main goal of the feature reduction procedure. The principal component analysis is used in this study to reduce the data set's dimensionality. By transforming an existing collection of parameters into a new set known as principal components (PC), PCA's primary goal is to shrink the size of data sets containing incorrect data. Indeed, the original data is linearly arranged on computers.

3.5.1.4 Similarity Measures

The distance between the attribute space and query image in feature vectors is used to determine how similar two brain MRI scans are. This work uses the equation to calculate the Manhattan distance between two places along axes at right angles.

$$d_{MD}(q,r_i) = \sum_{j=1}^{n} \left| f_j(q) - f_j(r_i) \right|$$

3.5.1.5 Classification Using Improved ELMPS

To improve the implementation of the classification of the new input image, an ELMPS-based tumor classification model is used. To reduce the uncertainty of ELM prediction using their input weights, this approach combines the ideas of ELM with density function and the Bayesian decision model. This model thus improves the classification's dependability. ELM is one of the fully trained neural networks and can learn properties from data. One hidden layer and one output layer are present. The output layer learns the mapping function, while the hidden layer learns the properties. With less network configuration, the ELM algorithm categorizes the data better. It is one of the more accurate, balanced, and time-efficient categorization techniques for certain classes. This technique achieves high accuracy, balance, and efficiency for certain classes despite having limited input-output data available, making it a valuable method in scenarios where data is scarce or expensive to obtain. Table 3.1 shows the confusion matrix of our model.

For model training, a limited number of input-output examples are used. Standard SLFNs with activation function g(x) and hidden nodes are statistically modeled as stated in Equation for N samples from the minimal spanning tree (xi, ti).

$$\sum_{i=1}^{N} \beta_i g_i(x_j) = \sum_{i=1}^{N} \beta_i g(w_i.x_j + b_i) = 0_j, j = 1,...,N$$

TABLE 3.1

Confusion Matrix for Performance Evaluation

Query Output Image	Tumor Image	Non-Tumor Image
Tumor Image	TP	FN
Non-Tumor Image	FP	TN

3.5.2 A DEEP BELIEF NETWORK WITH THE GREY WOLF OPTIMIZATION ALGORITHM

The brain tumor detection DBN-GWO-based classification model is a machine learning model that combines two different techniques, Deep Belief Networks (DBN) and Grey Wolf Optimization (GWO), for detecting brain tumors in medical imaging data. An FCM algorithm for image segmentation, a DBN that has been optimized for finding brain tumors in the input pictures, and two preprocessing techniques—contrast enhancement and skull stripping—are all included in the proposed model. Both the grey-level co-occurrence matrix (GLCM) and the grey-level run-length matrix (GRLM) features were extracted during the feature extraction phase. To improve the classification performance of the standard DBN, a grey wolf optimization (GWO) technique is used, as shown in Figure 3.4. The suggested strategy's effectiveness is carefully evaluated on the real-time data set. The effectiveness of the suggested strategies is shown by contrasting their implementation with other cutting-edge approaches in terms of performance measures. The input is an MRI

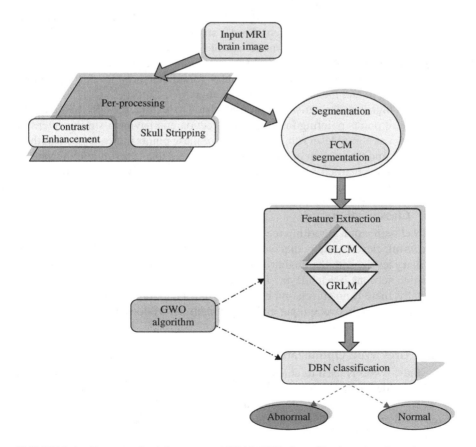

FIGURE 3.4 Framework of the proposed DBN-GWO–based brain tumor detection.

of the brain. The preprocessing, segmentation, feature extraction, and classification modules are crucial elements of this strategy.

3.5.2.1 Preprocessing

The contrast augmentation and skull stripping processes are carried out during this early stage of the suggested classification model.

3.5.2.2 Segmentation

An algorithm that is primarily utilized for picture segmentation is fuzzy means clustering. The output image IP is used as the input for FCM in this work. Segmentation is accomplished by splitting the picture spaces into various groups using the same image's pixel values. The typical FCM operates as given in Figure 3.4:

3.5.2.3 Feature Extraction

The process of feature extraction is a critical step in image processing techniques. In the suggested model, GLCM (grey level co-occurrence matrix) and GRLM (grey level run length matrix) features extracted from the input images are utilized for further analysis and classification purposes. From the input image, the grey co-occurrence matrix retrieves the second-order statistical texture information [4]. By generating GLCM, the suggested model establishes the correlation between pixels existing in the area of interest. The GLCM matrix includes the second-order joint conditional probability, co of one grey level y to another grey level z with inter-sample distance di and the given direction by angle $\downarrow$.

To extract higher-order statistical features related to texture, the GRLM method is used. A line of points pointing in a specific direction and having a similar intensity level is known as a grey-level run. The number of these points indicates the length of the grey level, and the number of occurrences is referred to as the run-length value. A run length is now regarded as a collection of nearby locations with comparable grey intensities in a given direction.

3.5.2.4 Classification

The DBN classification algorithm receives the designated features (F) as input. DBN is a successful, clever strategy that was first used in 1986 [13]. Typically, it has a few intermediary levels, with the final layer acting as a classifier. The input layer is made up of visible neurons in each layer. The output layer is made up of hidden neurons. Additionally, there is no correlation between visible and hidden neurons, yet there is a strong connection between input and hidden neurons. Hidden and visible neurons are connected symmetrically and exclusively.

3.5.2.5 Grey Wolf Optimization

In this study, GWO is used to adjust the retrieved features (F) in coordination with the DBN's hidden neurons. A well-known meta-heuristic algorithm, GWO describes grey wolves' leadership structure and hunting behavior [14, 15]. Grey wolves hunt in packs of five to twelve and cooperate to find their prey. The group's hierarchy is divided into four levels: (i) The alpha wolves, regarded as the pack leaders (either female or male). They can also decide when to go hunting, how long to travel, where to sleep, etc. (ii) The wolves in the next hierarchy, known as beta wolves, who assist alpha wolves in making decisions. (iii) The delta wolves are subordinates. (iv) The final level is omega wolves, also known as the scapegoats. All three, ß, serve as adjusting parameters in the GWO algorithm.

3.6   RESULT AND DISCUSSION

Deep belief networks are algorithms that generate outputs based on probabilities and unsupervised learning. They have both directed and undirected layers, and are made up of binary latent variables. In contrast to previous models, each layer learns the complete input in deep belief networks. In convolutional neural networks, the initial layers merely filter inputs for fundamental features like edges, while the subsequent layers integrate all the straightforward patterns identified by the earlier levels. On the other hand, deep belief networks operate globally and control each layer in turn.

An ELMPS-based classification model is implemented, in which the Wiener filter is employed to eliminate the unwanted pixels from the input image. A feature extraction method is used to extract relevant features based on the images' contour, boundary, and density. Then PCA is applied for feature reduction. Finally, the ELMPS algorithm is used to classify the images. A fast DBN-GWO–based classification model is developed with two preprocessing techniques, namely contrast enhancement and skull stripping, to enhance the image quality and FCM algorithm to segment input images. This model uses an optimized DBN to detect brain tumors in the input images. A GWO algorithm is applied to improve the classification performance of the conventional DBN.

3.7   SUMMARY

Brain tumor inference from brain MRI scans is a difficult endeavor due to distorted edges, complex organization of the brain, and external factors such as noise. To eliminate noise in the input image and improve the performance of the segmentation process, a hybrid clustering algorithm is developed. Generally, we can do preprocessing, feature extraction, and classification. Contrast enhancement and skull stripping are used in the preprocessing phase to remove noise and artifacts in the input image. Contrast enhancement is a pervasive method for amplifying the visible difference between adjacent pixels in an image by controlling contrast agents/media. Besides, contrast enhancement can also refer to features of anomalous skin lesions. The algorithm's stability can be increased through the clustering method, while minimizing the sensitivity of the parameters for achieving feature extraction to minimize the complexity and enhance the enactment of the classifier. In the end, a DBN is used for classification.

In the future, we can improve 3D photos, classify them more accurately, and create the optimization techniques that should be used. An efficient classifier is needed to group 3D objects into discrete feature classes so that brain illness diagnosis may be made using those characteristics. To improve and strengthen the implementation of the classification and segmentation of 3D images, additional research should focus on the preprocessing, feature extraction, segmentation, and classification stages.

The extraction and quantification of features from additional MRI modalities, including FLAIR, T1c-w, and T1-w slices, could be investigated to enhance the classification of neurological images into various classes of brain cancers, such as very small cell-like tumors, primary gliomas, metastases, grading of gliomas, lesions due to injury, and dementia. To achieve this, the features used in this study could be combined with those extracted from the additional MRI modalities. However, reducing the computational and time complexities is also crucial since medical diagnostic routines should take no more than a few minutes. Therefore, it is essential to optimize the system for rapid application on high-performance computing platforms that can run in parallel mode.

REFERENCES

1. Velliangiri, S., Anbarasu, V., Karthikeyan, P. and Anandaraj, S. P. (2022). Intelligent Personal Health Monitoring and Guidance Using Long Short-Term Memory. Journal of Mobile Multimedia, Vol. 18, No. 2, pp. 349–372.
2. Wang, L. and Healey, G. (1998). Using Zernike Moments for the Illumination and Geometry Invariant Classification of Multispectral Texture. IEEE Transactions on Image Processing, Vol. 7, No. 2, pp. 196–203.
3. Mirjalili, S. (2016). Dragonfly Algorithm: A New Meta-Heuristic Optimization Technique for Solving Single-Objective, Discrete, and Multi-Objective Problems. Neural Computing and Applications, Vol. 27, No. 4, pp. 1053–1073.
4. Harshavardhan, A., Babu, S. and Venugopal, T. (2017). Analysis of Feature Extraction Methods for the Classification of Brain Tumor Detection. International Journal of Pure and Applied Mathematics, Vol. 117, No. 7, pp. 147–155.
5. Kaur, G. and Oberoi, A. (2020). Novel Approach for Brain Tumor Detection Based on Naïve Bayes Classification. In Data Management, Analytics and Innovation; Singapore: Springer, pp. 451–462.
6. Zaw, H. T., Maneerat, N. and Win, K. Y. (2019). Brain Tumor Detection Based on Naïve Bayes Classification. 5th International Conference on Engineering, Applied Sciences and Technology (ICEAST), pp. 1–4, https://doi.org/10.1109/ICEAST.2019.8802562.
7. Arunkumar, N., Mohammed, M. A., Ghani, M. K. A., Ibrahim, D. A., Abdulhay, E., Ramirez-Gonzalez, G. and Albuquerque, V. H. C. (2019). K-Means Clustering and Neural Network for Object Detecting and Identifying Abnormality of Brain Tumor. Soft Computer, Vol. 23, pp. 9083–9096.
8. Badža, M. M. and Barjaktarović, M. Č. (2020). Classification of Brain Tumors from MRI Images Using a Convolutional Neural Network. Applied Sciences, Vol. 10, No. 6, https://doi.org/10.3390/app10061999.
9. Havaei, M., Davy, A., Warde-Farley, D., Biard, A., Courville, A., Bengio, Y., Pal, C., Jodoin, P.-M. and Larochelle, H. (2017). Brain Tumor Segmentation With Deep Neural Networks. Medical Image Analysis, Vol. 35, No. 1, pp. 18–31.
10. Gumaei, A., Hassan, M. M., Hassan, M. R., Alelaiwi, A. and Fortino, G. A. (2019). Hybrid Feature Extraction Method With Regularized Extreme Learning Machine for Brain Tumor Classification. IEEE Access, Vol. 7, pp. 36266–36273.
11. Bahadure, N. B., Ray, A. K. and Thethi, H. P. (2017). Image Analysis for MRI Based Brain Tumor Detection and Feature Extraction Using Biologically Inspired BWT and SVM. International Journal of Biomedical Imaging, Vol. 2017, pp. 1–12.
12. Bal, A., Banerjee, M., Chakrabarti, A. and Sharma, P. (2018). MRI Brain Tumor Segmentation and Analysis Using Rough-Fuzzy C-Means and Shape Based Properties. Journal of King Saud University - Computer and Information Sciences, Vol. 34, No. 2, pp.115–133.

13. Bauer, S., Wiest, R., Nolte-P, L. and Reyes, M. (2013). A Survey of MRI-Based Medical Image Analysis for Brain Tumor Studies. Physics in Medicine and Biology, Vol. 58, pp. 1–44.
14. Rajagopal, R., Karthikeyan, P., Menaka, E., Karunakaran, V. and Pon, H. (2023). Disease Analysis and Prediction Using Digital Twins and Big Data Analytics. In New Approaches to Data Analytics and Internet of Things Through Digital Twin; USA: IGI Global, pp. 98–114.
15. Karthikeyan, P. (2021). An Efficient Load Balancing Using Seven Stone Game Optimizations in Cloud Computing. Software: Practice and Experience, Vol. 51, No. 6, pp. 1242–1258.

4 Secured Data Dissemination in a Real-Time Healthcare System

Meera S
Sharmikha Sree R
Dinesh Kumar S
Kalpana R A
Valarmathi K

CONTENTS

4.1 INTRODUCTION

The primary point of this chapter is to get the progression of data through implanted systems (Internet of Things [IoT]) employed in medical services conditions utilizing Cryptography helps to ensure individual security, making smart devices smarter and reducing the effort required by users to perform key distribution tasks. By using cryptography, smart devices can securely exchange information without the need for users to manually distribute encryption keys. This makes the process more efficient and convenient for users.

IoT alludes to any actual item implanted with innovation fit for trading information. It is fixed to make a more proficient medical services framework concerning time, energy, and cost. By implanting IoT-empowered gadgets in clinical hardware, medical services experts want to screen patients more accurately – and utilize the information gathered from the gadgets to sort out who needs the most active consideration [1]. By capitalizing on this organization of gadgets, medical services experts could utilize information to arrange proactive administration. As is commonly said, counteraction is superior to the fix. In cryptography, each client has a secret key and distributes the public key to the next client. A client's secret key does not change over the long run unless the client has been compromised. Symmetric cryptography, such as AES (Advanced Encryption Standard), would not be efficient in this scenario because each set of clients would require a unique secret key.

Distributed computing is a web-based method that gives shared PC handling assets and information to PCs and different gadgets on request. It is a model for empowering universal, on-request admittance to a shared pool of configurable figuring assets (e.g., PC organizations, servers, capacity, applications, and administrations). A trust framework estimates client strategy infringement and keeps out miscreants. A cloud service provider can track user behavior by monitoring their interactions with the system and collecting data on their usage patterns. This information can be used to improve the service and provide personalized recommendations to users. However, it is important for providers to be transparent about their data collection practices and ensure that user privacy is protected. As the quantity of enrolled clients increases on the cloud, client-level trust assessment will be higher. Clients could be grouped because of advancing records to screen boundless [2].

Exergaming systems have the potential to improve the physical and cognitive health of older adults. By incorporating additional technologies such as an IoT-enabled glucose meter and XMPP messaging protocol, these systems can provide personalized feedback and support to help users better manage their health. Overall, these advancements in technology have the potential to greatly benefit the aging population. The proposed calculation consolidates a self-validating instrument to forestall ill-disposed control of message trades during the convention. The double-dealing of exact estimations defines quantization limits, and a heuristic log probability proportion gauge accomplishes a better mystery key age rate [3].

A pilot application, Focus Drive, has been created to exhibit security and protection insurance. Focus Drive fosters an application running on cellphones as a foundation application to screen the cellphone's speed progressively. Running this application, a PDA will naturally empower and debilitate the messaging capability as indicated by the driving velocity and street conditions [4]. A proposed pre-dissemination scheme could greatly improve an organization's flexibility compared to current plans. The scheme has a beneficial property where the likelihood of nodes other than those compromised being impacted is negligible when the number of compromised nodes is below the limit. This property reduces the initial impact of a smaller network break, improving the overall resilience of the system. By implementing this scheme, organizations can better protect against potential network disruptions and maintain consistent operation. [5].

4.2   SECURED DATA FRAMEWORK

In the available framework, wireless channel secret key extraction techniques are used. The process starts from a request message. After receiving the request message from BS (base station), it validates the sender. If the sender is valid, it takes the IP and port of destination for validation; if the receiver is also valid, then it adds random numbers to the request message. The request message and random number will be sent to the key generation process [6]. The generated key is forwarded to the key distribution process to send it to both the sender and receiver. Finally, the received key will be used by the sender and receiver for encryption and decryption of communicating messages.

Problem definition:
- Remote station key extraction procedures are not material through significant distance correspondence between a server and its clients.
- Key extraction procedures are inaccessible in a typical situation of the IoT. So key dissemination procedures are still essential to the security of the IoT by and large.
- Server or specialist is located in an emergency clinic, and clients or patients are receiving care remotely from their homes. This type of arrangement could enable medical professionals to remotely monitor patient health using telemedicine technologies and wearable medical devices, and provide timely interventions in emergency situations. Yet it is difficult to figure out a couple of encryption and unscrambling keys from various remote channels.

In the planned work, there is an original key dispersion approach for constant administration in the IoT. The checked clinical information for medical services is exceptionally connected with individual protection, which ought to be ordinarily communicated as code texts. In situations where only authorized information administrators are permitted to access sensitive data, requirements for real-time and high efficiency arise when a specialist is monitoring patients' medical issues remotely. In this scenario, the use of meeting keys for multiple clients becomes necessary to ensure secure communication and access control. The meeting keys allow the specialist to securely authenticate and communicate with each client, while also maintaining confidentiality and privacy of the sensitive data. This approach ensures that only authorized personnel can access the sensitive information and enables efficient remote monitoring and management of patients' medical conditions. A trusted outsider introduces the framework by creating a few boundaries [7]. These boundaries will be used to create public and confidential keys for all elaborate information administrators and clients. When another information director or client joins the framework, the trusted outsider processes and sends a confidential key to the new individual by a safe channel. The distribution of meeting keys enables efficient and secure communication between the parties involved, whether online or offline, and allows for seamless exchange of information in a secure and controlled manner.. Clients get the meeting keys from the comparing information directors for information assortment. They utilize these keys to scramble their own records and send them to the cloud.

The requirements, in particular, are a detailed specification of the necessary features and functionalities of software products. This initial step is critical to the software development process as it documents the specific requirements of a particular software system, including functional, performance, and security requirements. The requirements specification serves as a basis for further analysis, design, and implementation phases of software development, and ensures that the resulting product meets the needs and expectations of stakeholders. By clearly defining and documenting the software requirements, the development team can ensure that the software is developed to meet the desired quality standards and user need. The prerequisites likewise give usage situations from a client, a functional, and a managerial point of view. The reason for programming necessities in particular is to give an itemized outline of the paper, its boundaries, and objectives. This portrays the paper interest group and its UI, equipment, and programming necessities. It characterizes how the client, group, and crowd see the paper and its usefulness. Figure 4.1 describes the architecture diagram of the system.

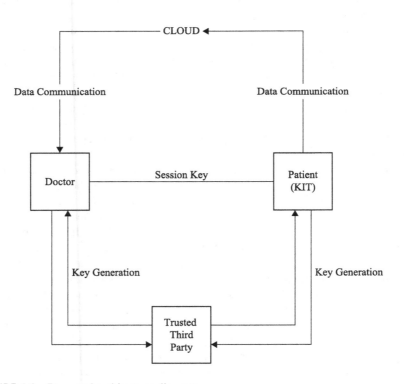

FIGURE 4.1 Proposed architecture diagram.

4.3   FRAMEWORK STRUCTURE

The framework has four components: (1) registration and key distribution, (2) advanced encryption standard key generation, (3) encrypted data dissemination via cloud, and (4) Raspberry Pi configuration. The functions are explained as follows.

4.3.1   REGISTRATION AND KEY DISTRIBUTION

The admin can log into the hospital website, register doctor details, and share the username and password. The admin has to register patient details. The server in turn stores the patient information in its database. The admin then selects a doctor based on the patient's disease. The trusted third party generates a public key and a private key for each doctor and patient. After generating the keys, the doctor and patient will receive the public and private keys.

4.3.2   ADVANCED ENCRYPTION STANDARD KEY GENERATION

The doctor (data manager) generates a session key for each patient. The data manager will receive the patient's public key from a third party. The patient's session key will be encrypted using this public key, and the encrypted session key will be sent to the patient. The session key will be sent to the patient using the RSA (Rivest–Shamir–Adleman) algorithm.

4.3.3   ENCRYPTED DATA DISSEMINATION VIA CLOUD

The patient's random data value will be sent to the doctor. There are two types of encryption process during the transformation. In the first process, the sensing data is encrypted using their own session key. In the second process, the encrypted data is again encrypted using the doctor's public key. Finally double encrypted sensing data will be sent to the doctor through the cloud [8–10].

4.3.4   RASPBERRY PI CONFIGURATION

The Raspberry Pi is configured to support the input data to be processed. The health status is sent to the doctor by encrypting the kit-connected data using the session key and again when encrypted using the doctor's public key. Whereas the doctor is online, the kit-connected data is moved to the cloud, and data is transferred to the doctor. In case the doctor status is offline, the data is stored in the cloud. When a doctor comes online, this data is automatically retrieved.

Figure 4.2 explains the sequence diagram of the framework. To secure healthcare data in the cloud using registration and key distribution, you can implement secure authentication methods such as two-factor authentication and biometric verification to ensure that only authorized users can access the data. Additionally, you can use a secure key management system to distribute and manage encryption keys for data protection. Advanced Encryption Standard (AES) key generation can be done using a variety of methods, such as using a secure random number generator

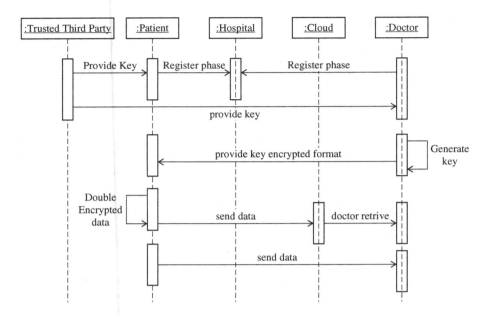

FIGURE 4.2   Sequence diagram.

to create a unique key for each user or using a key derivation function to generate
a key from a user's password. The key should be of suitable length and strength to
provide an appropriate level of security. To securely disseminate encrypted data
via the cloud, you can use a variety of encryption technologies such as Transport
Layer Security (TLS) or Secure Sockets Layer (SSL) to protect data as it is transmitted over the network. Additionally, you can use cloud-based encryption services to encrypt data at rest in the cloud. To configure a Raspberry Pi for secure
healthcare data storage, you can use various encryption tools such as dm-crypt,
LUKS, or TrueCrypt to encrypt the data stored on the device. Additionally, you
can use a firewall and intrusion detection/prevention system to secure the device
against unauthorized access. Secure protocols such as SFTP, SSH, and HTTPS can
also be used to secure the data transmission to and from the device. It's important
to note that securing healthcare data is a complex task that requires a multi-lay-
ered approach. It is essential to implement a comprehensive security strategy that
includes not only technical measures but also organizational and administrative
controls. It's also important to comply with relevant regulations and standards such
as HIPAA, HITECH, etc.

4.4   CODING PRINCIPLES

Coding principles are rules to programming that spotlight the actual design and
presence of the program. They make the code more straightforward to peruse, comprehend, and keep the information. This aspect of the framework really executes the
diagram created during the planning stage. The coding determination ought to be

FIGURE 4.3 Back-end server.

so that any developer should have the option to figure out the code and can achieve changes at whatever point is essential. Some of the standards required to achieve the aforementioned goals are as follows.:

- Naming conventions shows.
- Esteem shows.
- Content and remark technique.
- Message box design.
- Figure 4.3 depicts the back-end server.

Naming Conventions: Naming shows of classes, information, parts, systems, and so forth ought to be self-illustrative. One ought to try and get the significance and extent of the variable by its name. The shows are embraced for simple comprehension of the planned message by the client. So following the conventions is standard. These shows are as follows:

Class Names: Class names issue identical spaces, start with capital letters, and have blended cases.

Part Capability and Information Part Name: Part capability and information part name start with a lowercase letter, with each resulting letter of the new word capitalized and the remaining letters in lowercase. Figures 4.3 depict the overview of backend server. Figures 4.4 and 4.5 show the execution screenshots.

4.4.1 VALUE CONVENTIONS

- Esteem shows guarantee values for factors anytime of time. This includes the following:
  - Legitimate default values for the factors.
  - Legitimate approval of values in the field.
  - Legitimate documentation of banner qualities.

FIGURE 4.4   Front screen of the web page.

4.4.2 Script Writing and Commenting Standard

Script writing is where space is most significant. Contingent and circling explanations are appropriately adjusted to work with simple comprehension. Remarks are incorporated to limit the number of item that could happen while going through the code. Figures 4.5 and 4.6 show the execution screenshot that explains the sequence diagram of the framework.

FIGURE 4.5   Account creation page.

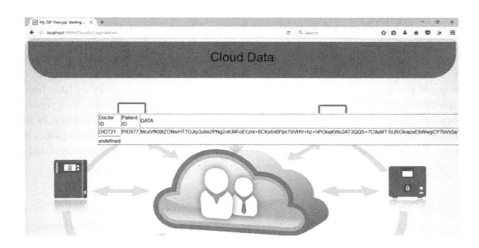

FIGURE 4.6 Cloud secure data.

4.4.3 MESSAGE BOX FORMAT

When something has to be prompted to the user, he or she must be able to understand it properly. To achieve this, a specific format has been adopted for displaying messages to the user. They are as follows:

- X – User has performed illegal operation.
- ! – Information to the user.

Testing is performed to recognize mistakes. It is utilized for quality confirmation. Testing is a basic piece of the whole turn of events and upkeep process. During the integration testing phase, the objective is to verify that the software component has been accurately and completely integrated into the system design, and to ensure the correctness of the overall design.. For example, it's important to identify and address any logic flaws in the design before coding begins. Otherwise, the cost of fixing these issues may be significantly higher later on. Identification of configuration shortcomings can be accomplished through examination as well as walkthrough.

Testing is one of the significant stages in the product advancement stage. The testing can significantly impact the effectiveness and efficiency of the software product being tested.

1. Static testing is utilized to explore the primary properties of the source code.
2. Dynamic testing is utilized to examine the behavior of the source code by executing the program on the test information.

4.5 SUMMARY

This chapter proposed an efficient key distribution for disseminating data with confidentiality and security. In this system, doctors can sensibly retrieve the health record of clients either online or offline. With the assistance of the cloud, the retrieval of personal records of patients can be done while guarding the identity of the clients simultaneously. Concerning security, we use the AES algorithm to generate session keys and disseminate compound individual random values as session keys to the mobile clients. This system provides a secure flow of data in IoT-enabled devices by preserving personal privacy bound to the smart devices. Future study may involve enhancement of additional features in smart devices.

REFERENCES

1. Robert S. H. Istepanian, Sijing Hu, Nada Y. Philip and Ala Sungoor. The Potential of Internet of M-health Things (m-IoT) for Non-invasive Glucose Level Sensing. In: 2011 Annual International Conference of the IEEE Engineering in Medicine and Biology Society, pp. 5264–5266, 2011.
2. Dijiang Huang, Zhibin Zhou, Le Xu, Tianyi Xing and Yunji Zhong. Secure Data Processing Framework for Mobile Cloud Computing. In: 2011 IEEE Conference on Computer Communications Workshops (INFOCOM WKSHPS), pp. 614–618, 2011.
3. Kim Thuat Nguyen, Maryline Laurent and Nouha Oualha. Survey on Secure Communication Protocols for the Internet of Things. Ad Hoc Networks, pp. 17–31, 32, 2015.
4. Charalampos Doukas, Ilias Maglogiannis, Vassiliki Koufi, Flora Malamateniou and George Vassilacopoulos. Enabling Data Protection Through PKI Encryption in IoT M-Health Devices. In: 2012 IEEE 12th International Conference on Bioinformatics & Bioengineering (BIBE), pp. 25–29, 2012.
5. Rodrigo Roman, Cristina Alcaraz, Javier Lopez and Nicolas Sklavos. Key Management Systems for Sensor Networks in the Context of the Internet of Things. Computers & Electrical Engineering 37(2), pp. 147–159, 2011.
6. Chunxuan Ye, Suhas Mathur, Alex Reznik, Yogendra Shah, Wade Trappe and Narayan B. Mandayam. Information-Theoretically Secret Key Generation for Fading Wireless Channels. IEEE Transactions on Information Forensics and Security 5(2), pp. 240–254, 2010.
7. Robert S. H. Istepanian, Emil Jovanov and Yuan Ting Zhang. Introduction to the Special Section on M-Health: Beyond Seamless Mobility and Global Wireless Health-Care Connectivity. IEEE Transactions on Information Technology in Biomedicine 8(4), pp. 405–414, 2004.
8. Sarveshwaran Velliangiri, R. Rajagopal and P. Karthikeyan. Trust Factor Based Key Distribution Protocol in Hybrid Cloud Environment. Scalable Computing: Practice and Experience 20(2), pp. 349–364, 2019.
9. Yuan Wang, Duncan S. Wong and Lliusheng Huang. One-Pass Key Establishment Model and Protocols for Wireless Roaming With User Anonymity. International Journal of Network Security 16(2), pp. 129–142, 2014.
10. Periyasami Karthikeyan and M. Chandrasekaran. Dynamic Programming Inspired Virtual Machine Instances Allocation in Cloud Computing. Journal of Computational and Theoretical Nanoscience 14(1), pp. 551–560, 2017.

5 Modality Classification of Human Emotions

Balasaheb H. Patil
Premanand K. Kadbe
Shashank D. Biradar
Trupti V. Bhandare
Parshuram N. Arotale

CONTENTS

5.1 INTRODUCTION

Emotions can be treated as a function that allows an organism to respond adaptively to environmental challenges rather than a subjective state. In recent years, there has been a debate about the nature of emotions. Emotions are the subjective state of consciousness. Emotions or emotional states are broadly categorized into various categories such as feedback, central nervous system, arousal, and cognitive theory. Emotional experiences are the result of previous emotional processes. The brain recognizes emotionally important events and produces responses to stimuli [1]. This requires feedback and awakening theory. Central and cognitive theories assume that the emotional experience is based on a preassessment of the situation, which further determines the content of the emotional experience [2]. Therefore, emotions are an important rating system that determines whether a particular situation is potentially harmful or beneficial to an individual. Fear is a basic human emotion that can trigger an unconscious assessment of potential danger or threats, leading to a conscious emotional experience. This process of unconscious assessment is shared by many animals and is independent of their evolutionary stage of development. This subjectively conscious experience of fear is the state of mind that occurs when the brain's defense system is activated, which detects threats and organizes appropriate responses. Therefore, fear and other emotions reflect the representation of evolutionary activity of the nervous system and the reactions they evoke as conscious content. Emotional behavior can be triggered by sensory inputs that pass the emotional experience [3]. Face analysis can be mathematically divided into four sub-problems: face recognition in a scene, face identification, analysis of its expression, and recognition of related emotions. Various facial features are used to perform this process, using different computational approaches [4]. These approaches are primarily based on psychophysical approaches, not neuro-physical data. Various studies have shown that cells selectively respond to areas of the temporal neo-cortex and specific faces of the amygdala.

Despite the fact that emotional processing can occur without conscious awareness, cognition often affects emotional processing as well as determines it. Many emotion theorists suggest that cognition contributes to emotions by shaping the subjective way of interpreting the world [5]. Such theorists try to characterize different emotions in terms of different types of preceding cognitive states. The goal is not only to identify the cognitive precursors of various emotions, but more generally to identify some systematic structural principles that govern the relationship between such states and emotions. There is considerable neurobiological evidence that emotional processing is actually affected by cognitive processing [6]. For example, brain structures involved in emotional processing, such as the amygdala, receive input from many cortical areas in addition to sub-cortical input. These inputs are from most, if not all, input modality, but from the secondary and tertiary sensory regions rather than the primary regions. In fact, the cortical areas projected onto the amygdala do not selectively respond to stimuli associated with strengthening, and such cortical areas do not process the influence of emotion itself and the emotional processing of the amygdala [7]. A more detailed breakdown of how cortex and cognitive processing affect the myriad of complex emotions described by psychologists has not yet been described at the neural level.

Given the behavioral and neuro-anatomical evidence that cognition normally plays an important role in emotion processing, a viable neuro-emotional network model of emotion processing must accept input from cognitive processing units [8]. Some existing neural network models provide input to the arousal unit or emotion learning system of units that are thought to be involved in cortical and cognitive processing. Manic-depressive-based helplessness and depression theory result from false monoamine regulation; delusions and hallucinations lead to a false pattern matching process. These theories and models provide an important first step in modeling the interaction between cognition and emotion. While such interactions have been extensively studied within the framework of cognition, future neural networks need to pay more attention to how cognition is shaped and evokes the various complex emotions people experience.

The central issue of emotion modeling is to distinguish between emotion-evoking and non-emotional situations and to correlate emotion-evoking situations with appropriate emotional states [9]. In some cases, it is important to determine the emotional importance of an emotionally evoked situation. Cognition often plays an important role in inducing and strengthening emotions. In vertebrate neuro-anatomical studies, the amygdala is often involved in the emotional processing of sensory stimuli. Some neurons in the amygdala respond to certain categories of reward-related stimuli, while others respond selectively to visual stimuli. The amygdala receives input indirectly from the sensory nucleus of the thalamus and from the cortex [10]. Some emotional processes can occur independently, while cognitive processes can affect or trigger other emotions. Higher cognitive function is not always necessary for the organism to respond to the emotional aspects of the stimulus. Some artificial intelligence models relate to inference of emotions and emotion-related states in the context of language comprehension systems, or in the context of inference systems, where inferences about people's estimation goals are made from the described emotions. Here, inferences are made to generate a plausible explanation for the described emotional state. For each modality, you can develop individual modality classifiers to analyze different human emotions. Compared to the unimodal approach, the multimodal system approach provides better results. Modality approaches can be used to study different types of emotions, including naturally occurring spontaneous, provocative, and behavioral emotions. Various emotions such as anger, despair, sadness, and imitation can be categorized by negative valence with predefined arousal, while interest, joy, and pride are assigned by positive valence happiness type. Researchers in experimental psychology, artificial intelligence, and cognitive science recognize mutual influence and cognition. Possible feature extraction methods include facial feature extraction, body feature extraction, and language feature extraction [11]. Facial feature extraction uses facial animation parameters based on the subject's neutral facial expression. Various feature points can be obtained from the frames extracted from the image sequence mask of the MPEG-4 format video. The feature point confidence level in estimating facial animation parameters provides an estimate of facial expression. Select threshold determination levels integrated data from various terms using the selected classifier.

5.2 LINGUISTIC EMOTIONS

Emotion recognition from language is always dependent on a person's characteristics, such as gender, age, socio-culture, and personality, but language is important in defining the mood of a person's current state of mind. Assessing the potential of the speaker's vocalization and visual speech parameters (speech speed, pitch average, pitch range, pitch variation, intensity, speech quality, intelligibility, etc.) has a significant impact on human emotions. For emotions like fear, anger, sadness, joy, contempt, and surprise, all of these criteria can be compared. The harmonics of the laryngeal wave depend on the parameters for producing a particular frequency [12]. The muscles of the larynx become tense as a result of the tightening of the vocal cords. A person's inner thoughts can be easily revealed through vocal utterances.

Therefore, you can express different gestures with the help of different body parts, such as hands, arms, and head.

Below are a few different emotion types, including typical emotion words. Good emotions come from assessing an event that is rated by goals and interests. Happiness means being satisfied with the desired event. Happy emotions are another kind of emotion, a subtype of happiness, where joy about events that are desirable to others means happiness, and joy about events that are undesired to others is joy. Other types of well-being include feelings based on the future. Happiness for future desirable events means hope; happiness for confirmed desirable events means satisfaction; and happiness for unidentified undesired events means relief. Recognizing one's praiseworthy behavior means pride and recognizing others' praiseworthy behavior when considering the attributional feelings that result from assessing behavior that is assessed in relation to criteria and values means praise [13]. The well-being cum attribution compound emotions include pleasure when one approves of one's own praiseworthy conduct, appreciation when one approves of someone else's praiseworthy behavior, and thankfulness when one approves of the connected desirable events. Attraction emotions are the result of evaluating objects based on personal preferences and tastes. These emotions arise when an individual is drawn to an appealing object, which is often associated with a feeling of liking. Feeling disgruntled about an adverse event means being unhappy. Feeling disgruntled about a desirable event for others means resentment. Feeling disgruntled about adverse event intended for others means pity. Feeling displeased about a prospective undesirable event means fear. Feeling disgruntled about a confirmed undesirable event means having fears confirmed. Feeling disgruntled about an unconfirmed desirable event means disappointment. Disapproving of one's own inappropriate behavior means shame. Disapproving of another's inappropriate behavior means reproach. Disapproving of one's own inappropriate behavior and feeling displeased about the related undesirable event means remorse. Disapproving of another's inappropriate behavior and feeling disgruntled about the related undesirable event means anger. Being repelled by an unappealing object means dislike [14].

Regarding the impact of emotion on cognition, research on mood-congruent memory indicates that people in positive moods tend to recall positive material better than negative material, and vice versa for those in negative moods. On the other hand, research on affective state-dependent memory suggests that memory recall is better when the affective state during recall matches the affective state during learning. Both of these effects can be explained by neo-associations or connectionist models, where affective units activated during recall provide additional connections to the recalled material, thus increasing its accessibility.

Studies of the effects of emotional states on value judgment show all the relatively low levels of well-being caused by trivial well-being [15]. Finding time refers to making time for activities or experiences that bring joy, fulfillment, or satisfaction to an individual's life. This can include hobbies, spending time with loved ones, pursuing personal goals, or engaging in leisure activities. Finding time greatly enhances people's satisfaction with general life as well as specific aspects. Anxious individuals tend to anticipate fearful events more than angry individuals. Additionally, angry individuals tend to assign more blame to actors who perform actions that can induce anger compared to anxious individuals. [16]. Can the effect on both memory and judgment be explained by increased activation of mood-fitting materials? For example, people clearly rate happiness and life satisfaction more positively on sunny days than on rainy days, but not when they comment on the weather before making the rating. The best explanation for this is that external causes, i.e., emotions that cannot be consciously attributed to the weather, are implicitly used as relevant information to make decisions about other things. Neural network models can address these insights by using competitive binding schemes. When emotional arousal is associated with a possible cause, it cannot affect cognition or be associated with other possible causes. The effects affect cognitive processing at the level of the neurotransmitter system. Neurotransmitter that plays a role in the body's fight or flight response and is involved in regulating mood. In contrast, anxiety causes a dopamine-dependent activation system, primarily in the left hemisphere, which can affect working memory with a variety of distracting information.

The discovery of brain structures involved in emotional processing suggests how the neurotransmitter system affects cognitive processing. The activity of the amygdala can regulate the signal-to-noise ratio through direct projection onto the cortex. Amygdala cortical projections also elicit memory by providing depolarizing inputs to enhance long-term potentiation to encode significant stimuli, similar to the neural network model of reward-mediated learning described above. There is a possibility that epinephrine still affects memory. Neurophysiological evidence suggests how emotions affect cognitive processing. The receptive fields of neurons in the sensory area can be altered by sensitization or classical conditioning. Sensory neurons can be sensitized fairly commonly, but more interestingly, the receptive field properties of sensory neurons expand or shift toward reward or punishment-related stimuli, altering the sensory processing of stimuli. However, it is unclear whether these require input from such sensory systems as the amygdala. Finally, emotions can have a lasting effect on cognition through the release of hormones. The level of adrenal steroid harmonics that may help prevent depression is regulated by stress levels, which in turn regulates the density of hippocampal synaptic spines [17]. Therefore, emotion-related hormone levels can clearly affect memory function. Due to different mechanisms, different emotions can have different effects on perceptual processing. Emotions can affect cognition due to different neural network architectures. In the computer-assisted learning and mentoring (CALM) model, activating the excitation unit globally changes the weight of the connections between the modules, which improves the learning rate. This may be similar to increasing the availability of neurochemicals that promote learning. Alternatively, learn the sensation from the excitement unit to the unit that encodes the stimulus. Other emotion-driven free parameters of neural networks can affect the cognitive activity of the network. In the ART (adaptive resonance theory) model, the central neuro-modulatory parameters, which are said to resemble the locus coeruleus, are released when an unexpected emotional pattern is inputted to the network. This prolongs the transient activation caused by stimuli, and although there are some problems with this model, it explains some of the effects of inter-stimulator intervals on learning.

Neural network models use a variety of mechanisms to adjust the signal-to-noise ratio in ways that emotions can affect cognitive processing. In the CALM module, the activated excitation unit outputs random noise to all representation units in the module. The ART model contains a feedback mechanism that effectively enhances the activation of emotion-related expressions, which has advantages when competing for activation. Other neural network models for cognitive processing include attention mechanisms. While these mechanisms are interesting, the complete model necessarily includes a variety of mechanisms that can influence cognition independently or in combination. Next, the neural network model of the interaction between emotion processing and emotion recognition focused on the contribution of emotion processing to learning and memory [18]. Future neural networks need to address other aspects of the interaction between emotion and emotional cognition. For example, input from an active awakening system can selectively increase the learning rate of connections with entities participating in the presentation of rewarded or punished input. This allows you to shift or extend the expressive properties of these units, as well as the receptive fields of neurons that encode stimuli combined with rewards or punishments. A further challenge lies in the short-term and long-term effects of emotional hormonal effects on memory and cognition. Future models also need to expand the repertoire of emotional mechanisms and help determine if newly discovered biological mechanisms in animals can explain the behavioral effects of emotions observed in humans. However, in the long run, progress will rely heavily on models that make better use of the cognitive information needed to distinguish and strengthen more complex emotions.

Methods for detecting audiovisual emotions have started to catch the interest of researchers. Singing incorporates lip and jaw motions as visual language characteristics, with as ear, eyelid, and cheek movements, pitch, and energy as auditory features. In order to classify the extracted three-rivulet audiovisual features, a triple hidden Markov model (HMM) was created [19, 20]. The proposed method was tested on seven different emotional states (surprise, anger, joy, sadness, disgust, fear, and neutrality) and showed an 85% accuracy rate in detecting them. The audio and visual systems were built separately by De Silva and Ng. The nearest neighbor method was applied to classify audio systems once pitch was retrieved as a feature. While the HMM was being trained as a classifier, the video system used an optical flow method to track the movement of the corners of the mouth, brows, and lip margins. The findings of the classification of audio and video were combined under a rule-based framework.

The two main categories of cognitive neuroscientific techniques are (1) single-cell recording and (2) brain imaging. Single-cell recording has three primary drawbacks: (1) it is intrusive and requires brain surgery on the patient; (2) it is stressful and requires frequent dosing; and (3) it cannot be used during experimental operations. There is no time limit, and the fourth option is re-examination. Brain imaging technology essentially consists of positron emission tomography (PET), functional magnetic resonance imaging (fMRI), electroencephalography (EEG), magnetic electroencephalography (MEG), magnetic resonance imaging (MRI), and transcranial magnetic stimulation (TMS). It falls within the simulation category (TMS). The aforementioned techniques have the fundamental benefits of being rapid, accurate, noninvasive, and requiring no brain surgery. Blood flow can be indirectly measured using techniques like fMRI and PET. fMRI measures hemodynamic adjustment and is blood oxygen level dependent) (BOLD) [21]. The following are its drawbacks: central nervous system (CNS) blood flow levels are sensitive to artifacts (movement, cavities, changes in tissue impedance, etc.), which can change imaging features and reduce temporal resolution. The MEG's fundamental goal is to quantify the magnetic field that is produced outside the head as a result of the brain's normal electrical activity.

This enhances the EEG's spatial resolution and is hardly ever used in therapeutic settings. TMS, however, has very distinct advantages over the earlier technique. It temporarily impairs brain function using electromagnetic induction. Although this method is very concentrated and offers excellent temporal resolution, it can give rise to seizures in people's brains. Unfortunately, the procedure described above involves a complicated apparatus that can only be used in specific locations. Additionally, due to their lengthy latency, blood flow measuring techniques are not suitable for interactions. The large scanner and the procedure's delayed vascular response to local reactions are its principal drawbacks. Most significantly, it restricts the user's mobility, and the oxygen cycle in the brain can influence imaging. In this study, we will look at fundamental and basic issues and issues based on the EEG signals' interpretation of emotions.

Emotions and their expressions are used in a variety of mechanisms, such as signals, attention, motivational and interaction control, situational assessment, self-image and image-building of others, expectation formation, and inter-subjectivity. It plays a crucial role in decision-making, problem-solving, communication, negotiation, and adaptation to uncertain circumstances, in addition to intimately interfering neurologically with the mechanisms involved in cognition. Emotions consist of inner feelings and thoughts, as well as other inner processes that may go unnoticed by those experiencing the feelings. Individual emotional states can be influenced by the nature of the situation. Individual differences in apparent subjective emotional responses to a given stimuli are possible. Recent advancements in neuroscience, psychology, and cognitive science point to the surprisingly significant role that emotions play in irrational and intellectual conduct. When an individual is in a positive emotional state, their perception tends to be biased toward positive events, and the same is true for negative emotions. Emotional states can also impact decision-making processes. In the case of reading a text, experiencing a negative emotional state can lead to a negative interpretation.

These are categorized as different types of emotions that can be caused by different physiological signals from a person. Additionally, we can discuss the various emotional subtypes, the nature of feature extraction techniques, ways for evoking emotions, and physiological signals used to categories emotions. Human emotions fall into three different types: Motivational, basic and confident, or social. Pattern recognition was used to combine data from various subjects with different measurement parameters quickly, with a focus on t-tests, comparisons of ANOVA (analysis of variance), and finding physiological correlation attempts. Rosalind W. Picard researchers classified the physiological patterns of a group of six emotions (happy, sorrow, disgust, fear, joy, and anger) by presenting to the participant a video clip. Heart rate, body temperature, and skin conductance are the characteristics used for classification. This task proposes independent emotion detection (multimodal) by

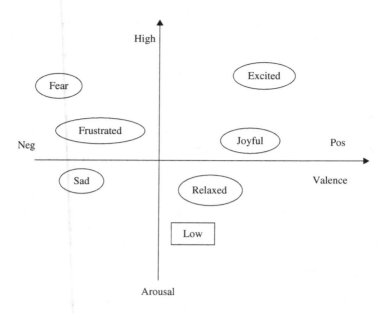

FIGURE 5.1   Discussion model.

collecting data from multiple subjects, and uses neural networks and support vector machines for simple emotions such as pleasure and discomfort.

According to P. J. Lang [22], one of the researchers, emotions can be classified according to their estimated valence (joy or unpleasantness) and arousal (calm or arousal). The degree to which arousal and valence are related is depicted in Figure 5.1. A link between physiological signals and arousal/valence is established when the autonomic nervous system is activated in response to emotional triggers. A person promptly and spontaneously experiences physiological changes in response to a stimulating event, and this response to these changes is referred to as emotion (William James [23]).

Therefore, the limbic system is engaged and plays a significant role in cognitive processes, EEG emotions, a central nervous system indicator, appear to be widespread and effective. The amygdala is located in various areas of the brain and plays an important role in detecting anxiety. Figure 5.2 shows the area of the brain where the amygdala resides.

Previous research, including animal experiments, has demonstrated that the amygdala is critical for perceiving fear in an animal's brain. Rats and monkeys are thought to behave less aggressively and fearfully when there is a distance between the two sides. As a result, both the experience of dread of oneself and the perception of fear of others are diminished by bilateral amygdala lesions. It merely lessens the sense of other people's dread. Damage to both sides of the amygdala makes it difficult to read negative emotions in facial expressions. Damage to both amygdala usually does not impair the perception of emotionally complex and static visual stimuli, except when these stimuli are accompanied by facial expressions as cues.

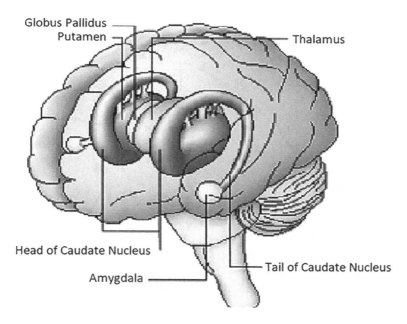

Globus Pallidus
Putamen
Thalamus
Head of Caudate Nucleus
Tail of Caudate Nucleus
Amygdala

FIGURE 5.2   Brain model for emotion.

This finding is particularly striking with respect to fear, and its perception is often impaired after bilateral amygdala injury.

Facial expressions generally improve the recognition of negative emotions in most people, but individuals with bilateral amygdala damage benefit much less from including facial expressions in emotion recognition compared to other subject groups. When analyzing emotional facial expressions, the brain activates several regions that develop at different stages, and some structures may develop multiple times. When the amygdala receives subcortical early (<120 ms) and late (~170 ms) inputs from the temporal lobe, it plays a significant role in processing fear stimuli.

5.3   RECORDING OF EMOTIONAL STATES CHANGES THROUGH EEG

Signals measured by the central nervous system are used in psychology to show how psychological changes and emotions are related. Most research used a t-test or an ANOVA to compare variables. The extraction of emotions from the physiological data of small pattern recognition systems, such as the identification system, has not received much attention. The majority of currently used methods for comprehending emotions are based on one modality, such as static or moving facial photos or videos, PET, fMRI, or EEG. Functional near-infrared spectroscopy, also known as fNIRS, is affordable, simple to use, and useful in the quickly developing field of brain computer interface (BCI), particularly for monitoring brain cognition and emotional

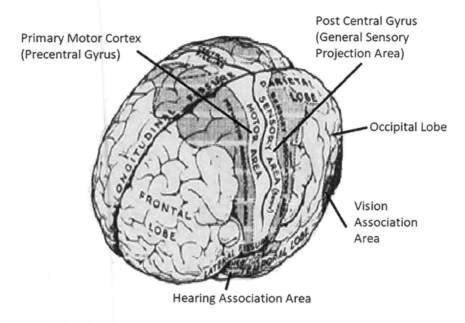

Primary Motor Cortex (Precentral Gyrus)

Post Central Gyrus (General Sensory Projection Area)

Occipital Lobe

Vision Association Area

Hearing Association Area

FIGURE 5.3 Motor areas and sensory areas of cerebral hemispheres.

states from the prefrontal cortex. fNIRS is used to track hemodynamic changes that occur during cognitive and emotional activities and to detect light flowing through cortical tissue. EEG neural activity estimate is the second modality used in this task. For more than 20 years, EEG has been used to examine people's emotional states. As long as a stable brain wave pattern is generated on the scalp, useful information about emotional states can be obtained. While fNIRS records hemodynamic activity in seconds, EEG recordings capture neuronal activity from the entire cortical surface in milliseconds.

Figure 5.3 shows the motor areas (toward the front) and the sensory areas of the cerebral hemispheres (toward the back). With respect to emotion-related EEG, researchers often focus on reducing activity in the alpha band (8 Hz to 13 Hz). Numerous studies indicate that the association between adult alpha activity and brain activation is inversely correlated.

The electrical variations between the human brain's resting and activating states in both hemispheres, as well as other physiological processes in reaction to stimuli, are recorded by EEG. EEG analysis was essential for investigations looking at brain asymmetry and emotions. According to this study (M. Lee, 2000 [24]), pointwise correlation dimension (PD2) analysis of the EEG data may or may not be used to determine happy and negative emotions. However, this PD2 represents a portion of the brain region's mental activity. In comparison to all other techniques, arithmetic activities are particularly effective at eliciting the emotions of the subject, and it can be deduced that more focus increases the complexity of the dynamics dimension of EEG measurements. It is the outcome of a person's cognitive evaluation of an experience and their physical reaction to it [25].

5.4 METHODOLOGY

Brain waves are used to find problems related to the electrical activity of the brain. The electrodes were used to detect various emotional stimuli such as valence, arousal, and predominance. You can analyze the complex dynamics of the human brain. Analysis of various brain activities determines the human condition. The signal is recorded electronically and analyzed accurately, but diagnosing the recorded signal requires a clinical diagnosis of one to two days. ANN (artificial neural networks) based EEG classification includes feature extraction algorithms, feature selection and fusion algorithms, and classification algorithms. The EEG signal from the headset is preprocessed and then sent to the feature extraction algorithm. This algorithm extracts specific characteristics of each signal from signals from a large number of sampled signals accessed by the headset. However, the size of the extracted feature vector is clearly smaller than the dimension of the original signal. The next step in feature selection is to extract meaningful features from the feature bundle from the previous step. The extracted features can be fused using transformations into a small and meaningful feature array for input. Once the array is formed, all inputs proceed to the next step in the classification algorithm. The output of the classifier diagnoses or detects neuropathy or normal human brain behavior. Analysis of the time or frequency domain can be used with feature extraction algorithms. The autoregressive (AR) model is a common algorithm used to classify EEG signals. Here, the essential characteristics of the EEG signal are modeled using the AR model. This creates a functional array of instantaneous EEG signals and further applies to ANN-based classifiers. The wavelet transform decomposes each signal into scale numbers to accurately identify the temporal characteristics of the signal in the time and frequency domains. For analysis with discrete wavelet transform (DWT), you can use filters with different cutoff frequencies.

Appropriate wavelet and resolution level selection criteria determine the efficiency of the wavelet coefficients extracted for classification. DWT(Discrete Wavelet Transform) has the ability to identify transient features of EEG signals in the time-frequency domain, which helps detect the elliptical confiscation of the wavelet packet decomposition (WPT), which decomposes both interleave level details and approximations. This gives you more flexibility in signal coding in several ways. Based on time-evolution estimations of nearby points in the state space or the local Jacobian matrix, the Lyapunov exponent employs dynamic quantitative measures of the signal to assess the stability of steady-state functioning. Approximate entropy represents the quantitative consistency of time series data.

5.5 EEG DATA SETS

Datasets are recorded by various subjects. Experiments with these recorded signals have been performed by many researchers who have proven that the results are moderate and useful for detection. We also certified experiments, set benchmarks, and shared resources used online. The key principle of this study is to seek modality of detectable stimuli and to quantify and manipulate them using an optimized number of electrodes that are effectively localized to the area responsible for audio-video processing. A test protocol is applied to the stimulus set, and signal changes in the brain are recorded. Filtering this signal removed the noise, performed an analysis on the signal, and calculated the spectrum of the signal. Then the classifier is used. Natural choices based on various basic emotions such as anger, joy, fear, curiosity, sadness, surprise, disgust, and acceptance are ways to detect emotions, but in this study, emotional valence and arousal were used as the primary factors for emotion detection. This method of choice has advantages in terms of suitability and simplicity over natural selection.

The voltage (signal) changes in response to the excitement of localized neurons. The release of sodium ions in the cell creates a potential difference inside the cell relative to the outside of the cell. At a specific threshold, sodium and potassium ions generate an electric potential. This creates an electrical signal that is transmitted through the dendrites to nearby neurons in a few milliseconds by changing the potential from -50 mV to $+20$ mV. In EEG, clear activity can be observed with electrodes located near the cortex. Changes in the amplitude and frequency of cortical activity in different states of consciousness can be easily classified. The 10–20 system is a standardized setup of electrode placement. Awakening is the ratio of alpha to beta and is used to feel relaxed and alert.

5.6 EXPERIMENTATION

An open-source, 3D-printed headset called Ultracortex can be used with any OpenBCI board. You can capture study-grade heart, muscular, and brain activity on an EEG. The headset receives EEG signals and samples up to 16 different EEG channels in 10 to 20 different locations. With the advantages of customization, comfort, and a 3D-printable headset, the Ultracortex Mark IV used in this study is the most recent version of Ultracortex and is compatible with all OpenBCI boards. The cutting-edge construction of Ultracortex uses a dry EEG sensor to shorten installation and walking distances. The Ultracortex Mark IV can target 35 different electrode positions in the 10–20 system, especially the motor cortex and visual cortex, as shown in Figure 5.4.

Figure 5.4 shows an Ultracortex Mark IV, based on the widely used 10–20 technique for placing electrodes. It has a very simplified assembly and is adjustable with the security of the electrode wiring to reduce noise and improve the aesthetics. You can put on the headset and receive a data stream of up to 16 EEG channels in less than 30 seconds after configuring the electrode holders on various frame nodes.

The subjects used were of different ages and genders. Subjects were asked to imagine expected emotions by advising them to think about past events in life. Another way to evoke emotions is to create audiovisual effects. Subjects were informed of what to do and what not to do during various tests. The selected subjects were of both genders, with undergraduates in their 20s and faculty members in their 50s. At the end of each stimulus, the subject was asked to relax or remain neutral. The data was recorded at different dates and times in multiple sessions on each topic. The recorded signal was first filtered with a bandpass filter. The bandwidths selected for these signals are 8–12 Hz and 12–30 Hz, the alpha and beta bands of excitation and valence, respectively. This reduced the amount of noise that can occur in a relaxed

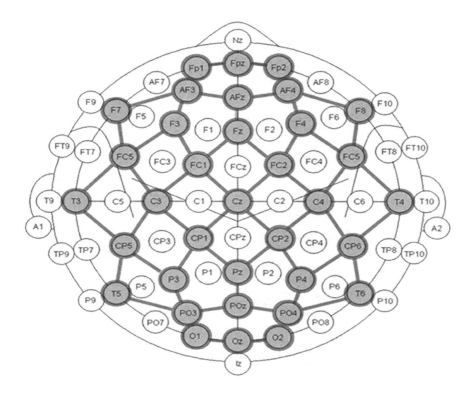

FIGURE 5.4   Ultracortex mark IV electrode spacing.

position. Wavelet transforms were performed on these filtered signals, and then principal component analysis was performed for feature extraction purposes. The Fisher Binary Linear Discriminant Analysis classifier was trained by emotion class.

An essential component of human-computer interaction is emotion detection. This is so that the advantages are clearer and more attainable. Figure 5.5 depicts the interface between the brain and computer. These days, the idea of introducing emotions into computers is gaining popularity. In many ways, the intuitive human-computer interface's frontier of emotions is one of the most unexplored.

5.7   CONCLUSION

Modality, arousal, and valence classifications based on alpha and beta band performance were observed with a classification rate of 85.6%. Obviously, the goal was to detect emotions rather than modality, making it more difficult to classify modality. The modality classification rate is 82% and is achieved by optimized selection of principal components. As shown in the results, the categories of arousal and valence were ranked higher than modality. The results obtained are close to previous studies in this area but can be improved by choosing the appropriate electrode positions and classification methods.

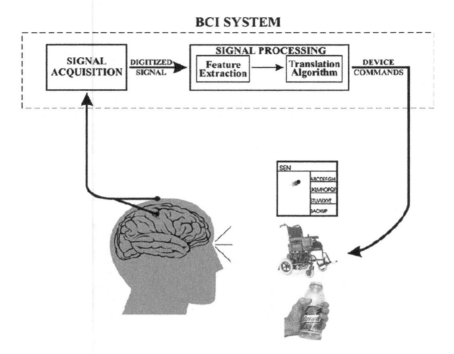

FIGURE 5.5   Brain computer interface system.

5.7.1 FUTURE SCOPE

Repositioning the electrodes can provide different functional signals for emotion recognition. The system can also be used for modality detection by placing appropriate electrodes in the visual and auditory cortex. Performance is also improved with large datasets.

5.8 SUMMARY

Human emotions and BCI are incredibly important concerns in electronics and the medical field. Wide research is still going on in these and other fields. Analysis of human emotions with different surrounding conditions, including environment and other factors, should be done to study the behavioral aspect of a person and perform continuous monitoring. On the basis of classification of human emotions, the researchers are analyzing the nature of a person and predicting future behavior, although there might be certain changes in surrounding conditions.

REFERENCES

1. Diano, M., Celeghin, A., Bagnis, A. and Tamietto, M. (2017). Amygdala response to emotional stimuli without awareness: Facts and interpretations. Frontiers in Psychology, 7, p. 2029.

2. Lindquist, K. A., MacCormack, J. K. and Shablack, H. (2015). The role of language in emotion: Predictions from psychological constructionism. Frontiers in Psychology, 6. doi:10.3389/fpsyg.2015.00444.

3. Tyng, C. M., Amin, H. U., Saad, M. N. M. and Malik, A. S. (2017). The influences of emotion on learning and memory. Frontiers in Psychology, 8. doi:10.3389/fpsyg.2017.01454.

4. Wójcik, W., Gromaszek, K. and Junisbekov, M. (2016). Face Recognition: Issues, Methods and Alternative Applications, In: S. Ramakrishnan (ed.), Face Recognition – Semi Supervised Classification, Subspace Projection and Evaluation Methods, IntechOpen, London. doi:10.5772/62950.

5. Izard, C. E. (2009). Emotion theory and research: Highlights, unanswered questions, and emerging issues. Annual Review of Psychology, 60, pp. 1–25. doi:10.1146/annurev.psych.60.110707.163539.

6. Velliangiri, S., Anbarasu, V., Karthikeyan, P. and Anandaraj, S. P. (2022). Intelligent personal health monitoring and guidance using long short-term memory. Journal of Mobile Multimedia, 18(2), pp. 349–372.

7. Pessoa, L. (2010). Emotion and cognition and the amygdala: From "what is it?" to "what's to be done?" Neuropsychologia, 48(12), pp. 3416–3429. doi:10.1016/j.neuropsychologia.2010.06.038.

8. Pessoa, L. (2017 May). A network model of the emotional brain. Trends in Cognitive Sciences, 21(5), pp. 357–371. doi:10.1016/j.tics.2017.03.002.

9. Gu, S., Wang, F., Patel, N. P., Bourgeois, J. A. and Huang, J. H. (2019). A model for basic emotions using observations of behavior in drosophila. Frontiers in Psychology, 10. doi:10.3389/fpsyg.2019.00781.

10. Matta, R., Choleris, E. and Kavaliers, M. (2020). Amygdala. In: V. Zeigler-Hill and T. K. Shackelford (eds.), Encyclopedia of Personality and Individual Differences, Springer, Cham. doi:10.1007/978-3-319-24612-3_726.

11. Wu, W. and Liu, T. (2023). Hossam Haick, electronic nose sensors for healthcare. In: R. Narayan (ed.), Encyclopedia of Sensors and Biosensors (First Edition), Elsevier, pp. 728–741. doi:10.1016/B978-0-12-822548-6.00097-2.

12. Zhang, Z. (2016 Oct). Mechanics of human voice production and control. Journal of the Acoustical Society of America, 140(4), p. 2614. doi:10.1121/1.4964509.

13. Xing, S., Gao, X., Jiang, Y., Archer, M. and Liu, X. (2018). Effects of ability and effort praise on children's failure attribution, self-handicapping, and performance. Frontiers in Psychology, 9. doi:10.3389/fpsyg.2018.01883.

14. Schindler, I., Hosoya, G., Menninghaus, W., Beermann, U., Wagner, V, Eid, M. and Scherer, K. R. (2017 Jun). Measuring aesthetic emotions: A review of the literature and a new assessment tool. PLoS One, 12(6), p.e0178899. doi:10.1371/journal.pone.0178899. PMID: 28582467; PMCID: PMC5459466.

15. Ruggeri, K., Garcia-Garzon, E. and Maguire, Á. et al. (2020). Well-being is more than happiness and life satisfaction: A multidimensional analysis of 21 countries. Health and Quality of Life Outcomes, 18, p. 192. doi:10.1186/s12955-020-01423-y.

16. Ashwin, C., Holas, P., Broadhurst, S., Kokoszka, A., Georgiou, G. A. and Fox, E. (2012 Mar). Enhanced anger superiority effect in generalized anxiety disorder and panic disorder. Journal of Anxiety Disorders, 26(2), p. 329–36. doi:10.1016/j.janxdis.2011.11.010.

17. Duman, R. S. and Aghajanian, G. K. (2012 Oct 5). Synaptic dysfunction in depression: Potential therapeutic targets. Science, 338(6103), pp. 68–72. doi:10.1126/science.1222939.

18. Tyng, M., Chai Amin, H., Ullah, M., Saad, M. and Naufal Malik, A. (2017). The influences of emotion on learning and memory. Frontiers in Psychology, 8, p. 1454. doi:10.3389/fpsyg.2017.01454.

19. Lu, C., Drew, M. S. and Au, J. (2001). Classification of summarized videos using hidden Markov models on compressed chromaticity signatures. Proceedings of the Ninth ACM International Conference on Multimedia – MULTIMEDIA '01. doi:10.1145/500141.500217.

20. Sarveshwaran, V., Joseph, I. T., Maravarman, M. and Karthikeyan, P. (2022). Investigation on human activity recognition using deep learning. Procedia Computer Science, 204, pp. 73–80.

21. Weintraub, A. and Whyte, J. (2016). Blood oxygen level dependent (BOLD). In: J. Kreutzer, J. DeLuca and B. Caplan (eds.), Encyclopedia of Clinical Neuropsychology, Springer, Cham. doi:10.1007/978-3-319-56782-2_10-3.

22. Lang, P. J. (1995). The emotion probe: Studies of motivation and attention. American Psychologist, 50(5), 372–385.

23. James, W. (1884). What is an emotion? Mind, 9(34), 188–205.

24. Lee, M., Lee, K. R., Kim, Y. H., and Choi, K. S. (2000). EEG-based emotion recognition in the valence-arousal space using a hybrid method of entropy-based feature selection and machine learning. International Journal of Human-Computer Interaction, 14(2), pp. 245–258.

25. Rajagopal, R., Karthikeyan, P., Menaka, E., Karunakaran, V. and Pon, H. (2023). Disease analysis and prediction using digital twins and big data analytics. In: New Approaches to Data Analytics and Internet of Things through Digital Twin, IGI Global, pp. 98–114.

6 COVID-19 Alert Zone Detection Using OpenCV

Sharmikha Sree R
Meera S

CONTENTS

6.1 INTRODUCTION

The coronavirus-2 extreme acute respiratory syndrome is the viral disease that causes the coronavirus. The World Health Organization (WHO) reports that as of October 18, 2020, there have been over 40 million illnesses and 1.1 million deaths reported globally, with more than 2.4 million new cases and 36,000 new deaths reported in the previous week. The virus is transmitted mostly between people after intimate contact, including microscopic droplets produced by coughing or sneezing. The infection is most hazardous during the first three days. Several common symptoms include weariness, a dry cough, and nausea. Serious and negative human repercussions have caused a global halt. Maintaining physical distance from the affected area

DOI: 10.1201/9781003345411-6

is necessary to prevent the disease's spread. When discussing the disease's prevention, a number of elements are at play. To mention a few, there is social distancing, face mask detection, posture detection, and crowd counting. These elements aid in figuring out what a person might be doing wrong and what might be done to protect themselves from the disease. Keep a space of two meters between any two people to remain secure. Face mask detection determines whether someone is wearing a mask or not and whether the mask has been worn properly. Our research suggests a way to tell if someone is wearing a mask. We also incorporate a version for crowd monitoring, which would alert the region's administrator if the number of people in the area exceeded a certain threshold [1, 2].

Even though wearing a mask is needed in densely populated areas, the vast majority of individuals in India have been unmasked. Most people do not use masks and do not practice social distancing in frequented areas such as marketplaces, bus stops, and train stations, resulting in a rise in the spread of COVID-19. In order to measure the proportion of persons wearing masks compared to those who don't in crowded areas like markets or a train stations using a video camera, our team is developing a deep learning model. Due to disruptions in health service delivery and normal vaccines, fewer individuals seeking care, and budget shortages for non–COVID-19-related diseases, the pandemic is likely to lead to increased fatalities from other causes. The second WHO pulse survey of 135 countries, conducted in March 2021, revealed widespread disruptions one year into the epidemic, with 90 percent of nations reporting one or more interruptions to key health services. Many individuals in India are still unaware of the dangers of COVID illness. Every day, 310 individuals in India die. Even though most people in India have been vaccinated, masks are necessary for populated areas because most people do not use masks and do not practice social distancing. Here YOLOv4 is utilized to recognize faces with and without masks. It employs CSP Darknet 53 as a backbone for feature extraction, whereas PANet is employed for feature aggregation and serves as the algorithm's neck. This project is delivered as extremely user-friendly software. We utilized the Python GUI library, and Tkinter provided the user interface. The interface allows users to give multiple forms of input for processing. We used NVIDIA for CPU and GPU computation. This gives improved performance by providing GPU utilization, GPU memory access and usage, power usage and temperatures, and time to the solution. They are a major element of today's artificial intelligence infrastructure, and new GPUs have been designed and tuned particularly for deep learning.

Using a live video feed and a deep learning mechanism, an object detection model, you can count the total number of persons wearing and not wearing masks. Additionally, the count will be kept in a database for later use. They suggest a skilled learning and computer vision-based strategy focused on the real-time robotized surveillance of people to locate unmasked faces in open spaces to create a secure atmosphere that leads to open safety. As a result, the suggested method benefits society by saving time and aids in slowing the coronavirus's spread. When the lockdown is lifted, bringing people together in public gatherings, shopping malls, etc., it can be executed successfully [3, 4].

In this deep learning mechanism, the model will check each person in the crowd, whether they were wearing a mask or not. If any one of them is detected not wearing a mask, his picture screenshot will be taken, cropped, and sent to a higher authority via mail or SMS through a communication medium to take action on the particular person. In a network of smart cities where closed-circuit television (CCTV) cameras watch over all public spaces, they suggest a system that limits the proliferation of COVID-19 by identifying individuals who are not wearing any facial masks. The municipal network alerts the appropriate authority when a person without a mask is found. A dataset of photos of people wearing and not wearing masks, gathered from multiple sources, was used to train a deep learning architecture. The trained architecture distinguished between persons wearing facial masks and those wearing none with 98.7% accuracy for never-before-seen test data. Our research could potentially help many nations worldwide by limiting the spread of this contagious disease [5, 6].

The YOLOv4 algorithm and deep learning method are utilized through a live feed camera to check various sorts of masks and moving people. This study attempts to create a face mask detector that recognizes various face masks. A YOLOv4 deep learning method has been selected as the mask detection algorithm to identify the face masks. The equipment has been placed at Politeknik Negeri Batam, State Polytechnic University, and the experimental results have been applied in real time. According to the experiment's findings, this device can accurately identify people wearing or not wearing face masks even while moving to different positions [4, 7, 8].

Computer vision applications can be created using TensorFlow and OpenCV. This task is enthusiastically suggested along with utilizing CCTV camera for every individual wearing or not wearing a face covering. If a particular individual isn't wearing a face covering, the CCTV camera takes a screen capture and edits his facial picture. His image is sent to match the dataset of the association to check his subtleties and his mail ID. An admonition message will be sent via email to the individual who was not wearing a covering. In this chapter, the authors propose a technique that uses YOLOVv V4 and OpenCV to detect whether individuals are wearing facial coverings. A bounding box is drawn around the individual's face to indicate whether they are wearing a mask or not. If an individual's face is stored in the database, the system identifies the person's name who is not wearing a facial covering and sends an email to them in advance, notifying them that they are not wearing a mask and urging them to take precautions [10, 11].

6.2   FRAMEWORK ENVIRONMENT

Face mask detection has two types of algorithms: YOLOv3 and YOLOv4. The camera can only recognize faces of people who are close to it. However, the current system is unable to determine the percentage of safety for a group of people, whether they are wearing masks or not. It is not feasible for a human operator to monitor the camera continuously to ensure that everyone in the area is wearing a face mask. The detection of face masks is in real time, which means it only shows whether a single person is wearing a mask. There is no user interface for jointly processing images, video, and real-time video processing.

6.2.1 CONSTRAINTS

6.2.1.1 No User-Friendly User Interface for Software

The biggest downside of face mask detection is there is no user interface for the software to perform data and store it for future use and real-time implementation of the live video through CCTV, mobile camera, webcam, etc.

6.2.1.2 Too Much Workload for Staff

Since it was almost impossible for human operators to sit in front of CCTV cameras to check people following social distancing, wearing face masks, and following the government rules for COVID protocols, the existing system does not help reduce the labor work.

6.2.1.3 Does Not Predict the Safety Percentage

The existing system does not predict the safety percentage of a number of people who are not wearing masks in crowded places. The camera recognizes people's faces, which is only near the camera.

6.2.1.4 Accuracy Decreases

The quality of the CCTV camera is very low, making it difficult to predict the face mask usage in real time, which becomes more problematic. Accuracy is not up to the level of the old algorithm model YOLOv3 used for detection in the existing system.

This project develops a user interface that takes the live video from the camera and takes the frame of the video. The frame of the video is analyzed for the faces in the photo. We count the total faces in the live video. For improving the accuracy of image processing, face mask detection is used. Video processing is used for face mask detection, frame by frame per second, and output is displayed frame by frame. For accuracy, it uses the YOLOv4 algorithm for face mask detection. For a user-friendly interface, develop software for user-friendly user interface to upload images, video, and real time simulation through the camera. In these faces, we find which faces are wearing a face mask and using that, we find the percentage of people wearing a mask. If it is above 75%, there is no need to alert the authorities. If it is less than 75%, we alert the concerned authorities to check that area. We create a pop-up message showing a warning to the user device to alert the particular authorities.

Advantages

- Workload reduced for the CCTV monitor staff.
- The accuracy of the prediction model increased.
- Easy to use.
- User-friendly interface.
- The processing time is relatedly less compared to the existing system.

6.3 FRAMEWORK STRUCTURE

Configuration designing arrangements with the different Unified Modeling Language (UML) charts for the execution of the venture. Configuration is a significant design portrayal of a thing that will be fabricated. Programming configuration is a cycle through which the necessities are converted into the portrayal of the product. Configuration is where quality is delivered in computer programming. Configuration is the resources to interpret client necessities into a completed item precisely.

6.4 ARCHITECTURE DIAGRAM

A system architecture is the conceptual model that defines a system's structure, behavior, and views. An architecture description is a formal description and representation of a system organized in a way that supports reasoning about the structures and behavior of the system. Figure 6.1 shows the framework's architecture, and Figure 6.2 shows the system execution plan.

A product application overall is carried out in the wake of exploring the total life cycle strategy for an undertaking. Different life cycle stages, such as requirement analysis, design, verification, testing, and implementation, result in a successful project outcome. Framework execution is a significant phase of a hypothetical plan that is transformed into a viable framework. Execution is the phase of the undertaking when the hypothetical plan is transformed into a functioning framework. Hence, it may very well be viewed as the most basic stage in accomplishing a fruitful new framework and giving the client certainty that it will work and be powerful. The execution stage includes cautious preparation, examination of the current framework and its limitations on execution, planning techniques to accomplish changeover, and assessment of changeover strategies. Each program is tried separately at the hour of advancement utilizing the information and it is confirmed that this program is connected in the manner determined in the project's detail. The PC framework and its current circumstance are tried per the client's general inclination. The framework that has been created is acknowledged and ends up being acceptable to the client; thus, the framework will be executed very soon. A straightforward working methodology is incorporated so the client can comprehend the various capabilities plainly and rapidly. In the final stage, the entire system is documented, which includes system components and operating procedures for the system administrator. The administrator must register with their details and provide a username and password. Patient

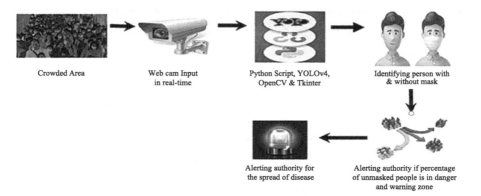

Crowded Area   Web cam Input   Python Script, YOLOv4,   Identifying person with
in real-time   OpenCV & Tkinter   & without mask

Alerting authority for   Alerting authority if percentage
the spread of disease   of unmasked people is in danger
and warning zone

FIGURE 6.1 Architecture diagram.

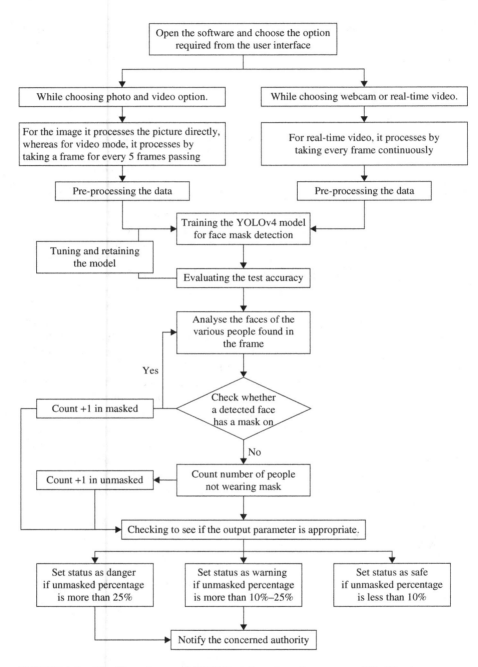

FIGURE 6.2   The Flow diagram of YOLOv4 detection of persons and Notifications.

details need to be registered by the administrator, which will be stored in the server's database. Based on the patient's illness, the administrator selects a suitable doctor [12]. The Trusted Third Party generates a public and private key for each doctor and patient. After the keys are generated, the doctor and patient will receive their public and private keys.

The implementation is divided into the following categories:
1. Creating user interface using Tkinter.
2. Processing photos through YOLOv4.
3. Screening Video through YOLOv4.
4. Enhancing the output parameter.
5. Implementation of real-time detection through the camera.

6.4.1 CREATING A USER INTERFACE USING TKINTER

We utilize the Tkinter module from a Python GUI to create the project's user interface. Tkinter is a typical graphical user interface tool. Tkinter is a Python binding tool provided with Microsoft Windows. We provide the user with access to the function they require. If they need to examine a single frame, they may use the upload image option; if they need to analyze video, they can use the upload video option; and if they need to check anything in real time, they can use the webcam option. The Tkinter.ttk module is used to style the Tkinter widgets. We utilize Tkinter.ttk to style Tkinter widgets in the same way that CSS is used to style HTML elements. We used a file dialog module for the file uploading section, and MessageBox Widget displays the message boxes in the applications. Figure 6.3 and Figure 6.4 show the front page of the framework design.

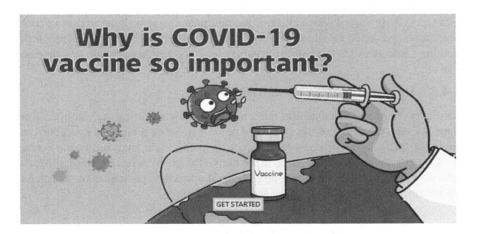

FIGURE 6.3 Awareness information page for the COVID-19 vaccination.

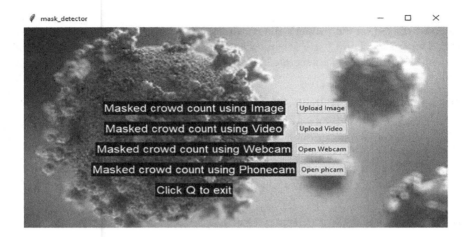

FIGURE 6.4 Homepage.

6.4.2 Processing Photos through YOLOv4

Once the file is uploaded using the user interface created in the previous module, it is processed by the YOLOv4 object detection algorithms. The GPU version employs the CSP Darknet 53 feature-extractor model. The YOLOv4's backbone is utilized for feature creation; in this case, it is used to detect the mask on the face. YOLOv4 selects PANet (path aggregation) for network feature aggregation. It is employed in the selection of the major feature from the backbone. This is done in YOLOv4's neck section. It finds the image's features via dense prediction. It is processed in the YOLOv4's head portion. In addition, the system employs both a bag of freebies and a bag of specials to enhance performance and extract novel features from the image. We receive the image with OpenCV, process it with YOLOv4, and return the image with the amount of confidence that is predicted for all faces detected in the image. If the percentage of people not wearing masks does not fall within the parameters, we alert the appropriate authorities.

6.4.3 Screening Video through YOLOv4

Similar to the image, this also uses the same process for analyzing the video received. Here, we take the frame of the video and process it. Since it is a vast procedure, we move the frame to five frames. Following processing, each analyzed frame is reassembled as a single video and saved in the output folder. We will alert the appropriate authority if the frame has a higher proportion of unmasked persons than specified in the parameter. Figure 6.5 shows the screening video.

6.4.4 Enhancing the Output Parameter

In this step, we examine the output of the analyzed frame or image to see if the percentage of unmasked people is greater than 25%. If this is the case, the status will be updated to danger, and we will notify the appropriate authorities. When it is 10–24%,

FIGURE 6.5   Image or video processing.

we just modify the status to a warning, so there's no need to alert the authorities. The status is changed to safety if it is less than 10%.

6.4.5   IMPLEMENTATION OF REAL-TIME DETECTION THROUGH THE CAMERA

In this step, we are trying to analyze the real-time camera for mask detection. It is similar to video processing, and it takes the live feed frame by frame and checks for masked and unmasked people. We are combining the device's webcam for real-time video and a phone camera to simulate the sense of a live camera analysis.

System testing verifies that the entire integrated software system meets requirements. It tests a configuration to ensure known and predictable outcomes. An example of system testing is the arrangement-oriented system mixing test. System testing is based on procedure similes and flows, emphasizing pre-driven process links and additional points. Figure 6.6 shows the output parameters of the system.

6.5   SUMMARY

The facial covering discovery module checks for individuals who are not wearing facial coverings appropriately and reports something very similar to the administrator. Taking the live camera feed, the model will count the number of individuals who are not wearing a facial covering or not wearing the face in a legitimate way, which is displayed to the administrator on the application. The count will be refreshed consistently to keep up with the exactness of the current circumstance for the administrator. For this model, we assess our methodology on the uniquely designed information base, which comprises around 500 pictures of the covered and exposed faces marked accordingly. We are characterizing the greatest clusters as 4,000. Sixteen subdivisions are shaped.

FIGURE 6.6 Output parameter (danger, warning, safe).

Each clump size has been characterized as 64. This way, 256,000 pictures are shaped. The group-counting model will count the quantity of individuals now present in a given region. The camera introduced from a bird's-eye view will catch the live feed of the area and will send it for examination, from which an ongoing count will be obtained, which will be displayed to the framework administrator through an application. The framework administrator can likewise set a limit as an incentive for the quantity of individuals nearby. The system will send an alert or a warning if the calculation exceeds the threshold value. The count will be refreshed consistently with the counting of every individual either entering or leaving the premises.

The group-counting model was likewise tried on a video transfer. It contains a virtual line that would work as an obstruction. The count would be expanded if the individual maneuvers upward (inside the room) and crosses the hindrance. If the individual maneuvers downward (outside the room) and crosses the obstruction, the count would be diminished. The number of individuals inside would likewise be shown.

6.6   FUTURE ENHANCEMENT

One way to address this issue in future improvements is by attempting to identify the rear face of the subject to determine if they are wearing a mask. However, in cases where the subject's face is not visible to the camera, it may be challenging to determine whether they are wearing a mask or not.

REFERENCES

1. P. N. Amin, S. S. Moghe, S. N. Prabhakar and C. M. Nehete, "Deep Learning Based Face Mask Detection and Crowd Counting," 2021 6th International Conference for Convergence in Technology (I2CT), 2021, pp. 1–5, doi:10.1109/I2CT51068.2021.9417826.

2. M. Rahman, M. H. Manik, M. Islam, S. Mahmud and J.-H. Kim," An Automated System to Limit COVID-19 Using Facial Mask Detection in Smart City Network," 2020 IEEE International IOT, Electronics and Mechatronics Conference (IEMTRONICS), 2020, pp. 1–5, doi:10.1109/IEMTRONICS51293.2020.9216386.

3. S. Susanto, F. A. Putra, R. Analia and I. K. L. N. Suciningtyas, "The Face Mask Detection For Preventing the Spread of COVID-19 at Politeknik Negeri Batam," 2020 3rd International Conference on Applied Engineering (ICAE), 2020, pp. 1–5, doi:10.1109/ICAE50557.2020.9350556.

4. J. Zhang, F. Han, Y. Chun and W. Chen, "A Novel Detection Framework About Conditions of Wearing Face Mask for Helping Control the Spread of COVID-19," IEEE Access, 9, pp. 42975–42984, 2021, doi: 10.1109/ACCESS.2021.3066538.

5. S. Velliangiri, V. Anbarasu, P. Karthikeyan and S. P. Anandaraj, "Intelligent Personal Health Monitoring and Guidance Using Long Short-Term Memory." Journal of Mobile Multimedia, 18(2), pp. 349–372, 2022.

6. H. Adusumalli, D. Kalyani, R. K. Sri, M. Pratapteja and P. V. R. D. P. Rao, "Face Mask Detection Using OpenCV," 2021 Third International Conference on Intelligent Communication Technologies and Virtual Mobile Networks (ICICV), 2021, pp. 1304–1309, doi: 10.1109/ICICV50876.2021.9388375.

7. A. Negi, P. Chauhan, K. Kumar and R. S. Rajput, "Face Mask Detection Classifier and Model Pruning with Keras-Surgeon," 2020 5th IEEE International Conference on Recent Advances and Innovations in Engineering (ICRAIE), 2020, pp. 1–6, doi:10.1109/ICRAIE51050.2020.9358337.

8. A. Sevugan, P. Karthikeyan, V. Sarveshwaran and R. Manoharan, "Optimized Navigation of Mobile Robots Based on Faster R-CNN in Wireless Sensor Network." International Journal of Sensors Wireless Communications and Control, 12(6), pp. 440–448, 2022.

9. M. W. A. Das and R. Basak, "Covid-19 Face Mask Detection Using TensorFlow, Keras and OpenCV." 2020 IEEE 17th India Council International Conference (INDICON), 2020, pp. 1–5, doi: 10.1109/INDICON49873.2020.9342585.

10. A. Jain, K. Gandhi, D. K. Ginoria and P. Karthikeyan, "Human Activity Recognition with Videos Using Deep Learning," 2021 International Conference on Forensics, Analytics, Big Data, Security (FABS), 2021, December, Vol. 1, pp. 1–5.

11. P. Karthikeyan and C. Tejasvini, Review of Movie Recommendation System, 2022 8th International Conference on Advanced Computing and Communication Systems (ICACCS), 2022, March, Vol. 1, pp. 1538–1543.

12. D. P. Rajan, D. Baswaraj, S. Velliangiri and P. Karthikeyan, Next Generations Data Science Application and its Platform, 2020 International Conference on Smart Electronics and Communication (ICOSEC), 2020, September, pp. 891–897. IEEE.

7 Feature Fusion Model for Heart Disease Diagnosis

Bhandare Trupti Vasantrao
Selvarani Rangasamy
Chetan J. Shelke

CONTENTS

7.1 INTRODUCTION

The real-time world is becoming more and more automated as new technology is developed. Different methods have been suggested to achieve improvement in the performance of the automation process in order to increase system performance. The most challenging task of automation is medical data processing. Diagnosis systems were designed to process medical data and make decisions that consider the available data, which is processed using a sophisticated algorithm. The automated heart disease diagnosis system developed recently has more focus on data processing. Recently, a number of strategies have been developed for diagnosing and early predicting heart disease using physiological markers identified by electrocardiogram (ECG) or Cleveland datasets. Cardiac activity is a measure of fast fluctuations in the heart's normal rhythm that are dependent on a person's physical or mental well-being. Depending on whether the heart is contracting or expanding, the electrical impulses in the heart's operation are monitored in ECG signals. One of the most frequently cited observations in heart disease diagnosis is measure of fast fluctuations in the heart's normal rhythm that are dependent on a person's physical or mental well-being. To give an accurate prognosis of the heart status, the ECG signal processing procedure calls for great precision in storage and computation. It has become more and more common to process ECG data in order to diagnose various cardiac problems, including coronary artery disease (CAD), arrhythmias, heart valve disorders, and more. Also, there is a chance of postoperative complications in cardiac surgery. To know these complications,

ECG processing is very helpful [1, 2]. Longer hospital stays that result in severe illness and higher treatment costs are also caused by post-diagnosis problems and diagnosis delays [3]. Heart disease is more common in adults [4], but cases in people between the ages of 40 and 50 have been rising quickly in recent years. According to published literature, numerous cardiac disease diagnoses employ ECG signals. It was described in [5] how to locate the QRS segment (where QRS stands for "ventricular complex of the electrocardiogram (ECG)" which represents the depolarization of the ventricles of the heart) using the square double difference method for R-peak detection for ECG analysis. Utilizing three steps of processing, this method was designed for feature extraction from ECG data. The sorting, thresholding, and approximate region-matching processes are presented. The R-peak detection was developed based on the R-peak values' magnitude differences. However, the decision performance has been constrained by external inference. Skala et al. [6] present a feature extraction method with signal denoising. This method suggested a soft thresholding strategy for processing the ECG signal. The wavelet-based method is used for the denoising process of both signal and feature extraction. The soft threshold is used in the processing of the spectral decomposition bands. The enhancement in feature selection is possible using the K-nearest neighbor algorithm, which is described in [7]. This method uses a classifier model for ECG signal analysis by detecting different QRS patterns. This method used a bandpass filter to select the feature vector, and the noise was filtered out. The feature vectors were extracted using a bandpass FIR filter [8]. The QRS pattern has six characteristics total that were retrieved using a sliding window method. Adam et al. [9] present a method for signal denoising that uses a modified Weibull distribution for ECG. This independent component analysis-based blind method for signal extraction from the source is defined. Hilbert transformations were employed in the derivation of line patterns. A normalized multi-derivative wavelet was utilized to extract the feature vector and process it for detection and denoising. The Nyman-Pearson classification was created in [10], while the extraction feature classification was developed using the Euclidean distance. In [11], time relationships are used to build feature selections, and wavelet-based transforms are built for feature extraction. Yildirim et al. [12] describes a feature extraction and selection method utilized to diagnose cardiac disease using P and T peaks. The method being discussed creates multiple classifications for the diagnosis of heart disease. Geometric feature extraction is described in [13]. These methods extract features from ECG signals based on structural representation. We used discrete one-dimensional (1D) wavelet processing to maintain structural changes. However, the noise effect of the derived features leads to misclassification. Several classifiers, including support vector machines (SVMs), stochastic neural networks (SNNs), and multilayer perceptron (MLP) back propagation networks, are described in the classification model [14–16]. Methods of artificial intelligence (AI) diagnose and detect cardiovascular disease early. Sujith et al. [17] and Sardar et al. [18] provide an overview of how AI is used in cardiology diagnosis applications. These methods considerably reduce baseline wandering, and classification is carried out using linear mapping. A brand-new, kernel-driven classifier model is introduced in [19]. SVMs and artificial neural networks are the best choices for classification. Working with these two models as a classifier is described by the author in [20, 21]. Similarly, the Cross WT is described in [22], where the correlation is derived from two different signals. In order to detect heart disease, some new approaches are introduced based on the entropy of the wavelet and the signature descriptors [23–25]. However, there are limitations due to the complexity of the function, the accuracy of the function selection, and the correct linking of the learned functions. New functional fusion technology that integrates the characteristics of discrete monitoring data with continuous ECG signal functions are being developed for new functional representations for the recognition of heart disease. Features are selected based on their importance or criticality so that the system will take minimum processing time and will have improved accuracy of detection. According to the list, the contribution is:

1. A novel feature fusion strategy was developed using regression and weighted clustering.
2. Using a deep neural network model and a feature-selected classifier model.
3. Integrated ECG feature vector with electronic record data.

Seven sections make up the remaining portion of this chapter.

Section 7.2 discusses the feature representation in heart disease diagnosis, Section 7.3 presents the current method for the feature representations. Section 7.4 discusses the proposed method. Section 7.5 discusses simulation results and conclusion is discussed in the Section7.6.

7.2   FEATURE REPRESENTATION IN HEART DISEASE DIAGNOSIS

Numerous factors influence how heart disorders are diagnosed. Many observations have an inter-relative structure. The estimation performance and accuracy can be enhanced by the proposed selected feature. The medical records, which are discrete in nature, and the continuously changing ECG signal are the two main elements employed in the diagnosis of cardiac disease. There are 14 dominating feature representations available for the observation of the discrete feature vector. They are listed as follows:

1. Age: The patient's age is given in years.
2. Gender: Indicate if the patient is male or female.
3. Chest pain (CP): Describe the patient's chest pain. Patient with a history of usual angina (chest pain in the past) (angina).
4. The patient's medical history is ineffective, but the current chest discomfort is diagnosed as atypical angina (abnang).
5. The patient has a painful condition known as non-anginal discomfort and experiences brief chest pain (notang).
6. Asymptomatic patients are those who have a disease but have a low risk of developing heart disease (asympt).

7. Trestbps: A patient's initial monitoring of blood pressure at rest position.
8. Chol: The cholesterol density of the patient, expressed in mg/dl.
9. Thalach: Maximum heart rate or heartbeats per minute of the patient is indicated.
10. Exang: Angina is a type of chest pain. The value of Exang represents the presence of angina during the exercise.
11. Oldpeak: segment of an electrocardiogram (ST) variation value after exercise recovery.
12. Slope: This word describes how the slope changes during exercise.
13. Ca: Lists the numbers of the main vessels.
14. Thal: Shows the presence of thalassemia, i.e., the blood disorder.
15. Num: This shows the class of patients based on the severity of the disease, which ranges from 0 to 4.

These UCI (University of California, Irvine) characteristics, which characterize the measured distinct coefficient of the automation system, are easily accessible in the Cleveland database (https://archive.ics.uci.edu/ml/machine-learning-databases/heart-disease/). In our earlier work [26], we offer a feature selection methodology for the Cleveland dataset's monitoring feature. This database consists of a total of 76 feature values, out of which 14 are considered for research work, as each of them is medically significant. The reason to consider these 14 as medically significant is as follows:

1. Age: With the risk of getting cardiovascular or heart problems roughly doubling with each decade of life, age is the most important factor and has a greater impact. In adolescence, coronary fatty streaks can start to develop. According to estimates, persons 65 and older make up 82% of coronary heart disease fatalities. The rate of stroke after age 55 doubles after every ten years.
2. Sex: Compared to premenopausal women, men are at high risk of heart disease. It has been stated that a woman at menopause is at the same risk level as a man, but the World Health Organization (WHO) and United Nations (UN) data oppose this. Diabetic females are at higher risk of heart disease than diabetic males.
3. Angina: A common type of chest pain brought on by a shortage of oxygen ironic the blood supply to the heart muscle. Patients may feel uneasy in this situation. They may feel some pressure or tightness in the chest. Additionally, they could experience pain in their jaw, neck, back, shoulders, or arms. Sometimes the pain from angina can resemble dyspepsia.
4. Resting blood pressure: Atherosclerotic artery disease can harm your heart's blood supply. Your risk is further increased when high blood pressure coexists with other illnesses such as diabetes, high cholesterol, or obesity.
5. Serum cholesterol: The most frequent cause of arterial constriction is low-concentration lipoprotein cholesterol, or "bad" low-density lipoprotein cholesterol (LDL).Increased blood fat levels called triglycerides, which are

connected to food, increase a patient's risk of suffering a heart attack. Low levels of high-concentration lipoprotein or "good" cholesterol can reduce the risk of having a heart attack high-density lipoprotein (HDL).

6. Fasting blood sugar: When the hormone insulin, which is secreted by your pancreas, is not produced in sufficient amounts or is not properly processed, then your possibility of heart attack rises when there is a rise in the body's blood sugar.

7. Resting ECG: According to the United States Preventative Services Taskforce (USPSTF), the risks of screening with a resting or exercise ECG are at least as large as the possible advantages for low-risk people for cardiovascular disease. There is currently inadequate information to determine the balance between screening's advantages and disadvantages for those at intermediate to high risk.

8. When the maximum heart rate is achieved, the risk of cardiovascular problems increases, similar to the risk associated with high blood pressure. Even a slight increase of 10 beats per minute in heart rate can raise the risk of cardiac death by more than 20%. This increase in risk is comparable to what would happen with a 10 mm Hg increase in systolic blood pressure.

9. Exercise-induced angina: A tight, gripping, or squeezing type of pain or discomfort, angina can range in intensity from mild to severe. Angina is typically felt in the center of the chest, although it can also cause discomfort in one or both shoulders, the back, neck, jaw, or arm. However, it is not typically felt in the hand. The types of angina are (i) pectoris angina (stable angina), (ii) unstable angina (prinzmetal variant angina), and (iii) microvascular angina.

10. Peak exercise ST segment: On a treadmill ECG stress test, an ST-segment depression that is horizontal or downward-sloping and smaller than 1 mm 60–80 ms past the J point is believed to be abnormal. Electrocardiograms (ECGs) that show upsloping ST-segment depression during exercise are often referred to as equivocal tests. Generally, flat or down-sloping ST-segment depression at a lower workload (measured in METs) or heart rate indicates a worse prognosis and a greater chance of multi-vessel disease. The extent of ST-segment despair is especially important, since a successful treadmill ECG stress test is associated with a longer revival following a stressful episode. Diagnosis of severe CAD is further indicated by a ST-segment altitude >1 mm, which commonly denotes transmural ischemia. Such patients are normally referred for coronary angiography right away.

Furthermore, a continuous ECG signal monitoring system was used to convert the measured continuous parameter into a discrete representation. ECG signals play an important role in heart disease detection. The standard lead system to trace ECG signals and the basic components of ECG are described in this section. An ECG reads the cardiac muscle impulse. A typical 12-lead ECG displays the electrical motion of the heart as traced by electrodes located on the body's surface.

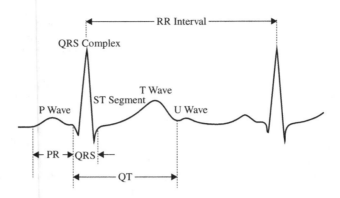

FIGURE 7.1   ECG signal form.

Figure 7.1 shows the standard 12-lead ECG waveform that illustrates the different intervals. Interval readings are useful to decide whether the working of the heart is normal or abnormal. Different intervals and their meanings are:

1. P wave: Left and right atriums continuously depolarize.
2. QRS complex: Reads depolarization of the right and left ventricles in the QRS complex.
3. ST-T wave: Ventricular repolarization.
4. U wave: A tiny "post-depolarization" event known as the U wave occurs at the start of diastole.
5. PR interval: The time between the commencement of the atrial depolarization (P wave) and the beginning of the ventricular muscle depolarization is known as the PR interval (QRS complex).
6. QRS duration: The length of time the ventricular muscle is depolarized (width of QRS complex).
7. QT interval: The length of time it takes for the heart to depolarize and repolarize.
8. PP interval: Atrial or sinus cycle speed.
9. RR Interval: Ventricular cycle velocity.

This illustrates how the heart works, and the rhythmic recurrence shows the variety of heartbeats. An average ECG signal is obtained throughout continuous monitoring, which may last between 24–48 hours. About 12 separate observations with variously located sensors can be employed for a detailed monitor. The ECG signals are buffered in 8–12-bit binary encoding and captured at sampling rates ranging from 125–500 Hz Data ranging from a few kilobytes to several megabytes can be stored in a file as discrete samples for processing. Interference during ECG capture or buffering adds extra overhead to the data that has already been buffered. In essence, an ECG signal is captured at a frequency of 0.05–100 Hz and a voltage of 1–10 mV. The variability in a heartbeat is represented by the five peaks and valleys that make up the ECG signal: P, Q, R, S, and T.

In a heart that is healthy, the variation of a P–R interval lasts about 0.20–0.22 seconds. The ECG feature interpretation includes different time intervals, such as PQRST time period and RR interval, and different peak values such as R peak, T wave, P wave values, etc. Based on the ECG interpretation results, various heart disorders, such as heart failure, CAD, heart arrhythmia, different heart valve diseases, or blockages are identified. The fluctuating coordinates of the signal define the ECG signal. The following are the 12 features used in this diagnosing process:

1. R pk cnt – R peak count in the signal.
2. Q Dur – Q duration.
3. R Dur – R-R interval duration.
4. R Ampl – R peak amplitude.
5. Q Ampl – Q peak amplitude.
6. S Ampl – R peak amplitude.
7. P Loc – P's location.
8. Q Loc – Q's location.
9. R Loc – R's location.
10. S Loc – S's location.
11. T Loc – T's location.
12. ST dev – ST deviation.

These characteristics can vary, which reflects the variation in heartbeats that are used to diagnose cardiac disease. These characteristics were read by the automation system designed for heart disease diagnostics as an input parameter in forming judgments and diagnosing various heart diseases. In a recent work [27], a feature fusion model for ECG characteristics and patient medical history is described. The cluster information gain for a normalized component serves as the foundation for the feature fusion model. The section that follows offers the suggested feature fusion methodology.

7.3   FEATURE FUSION METHOD

In order to accurately describe the changes in observational data, the representative aspects for diagnosis are crucial. Features for diagnosing heart disease are identified by capturing the sensor interface or by documenting a patient's physical or medical history. Because there are many sources of feature vectors, developing diverse processing and classification systems can be challenging and burdensome for automation systems. By combining various observations, the overhead in the automated process can be reduced. Fusion can be produced at the level of a feature, decision, or set of data. A feature-based fusion strategy aims to combine sensor monitoring features with patient record data [27, 28]. The feature fusion technique framework from [23] is depicted in Figure 7.2. The development of integration between the digitally recorded data and the physiological data measured by sensors.

Preprocessing the recorded data improves the quality of the signal representation. This includes signal normalization, filtering out missing data, and denoising. The proposed strategy was used to develop a feature fusion selection method based

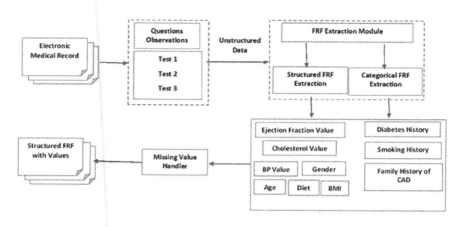

FIGURE 7.2 Feature fusion using a medical record [26].

on information gain and information entropy. An array of feature vectors that were gathered from sensors and data records are arranged here. The information gain factor (IGF) monitoring aids in feature selection optimization. The feature vector's IGF is shown as:

$$\text{IGF}(f,f') = v(f) - v(f,f') \tag{7.1}$$

The sensor's and the data's feature vector, respectively, are denoted by f, f'. The variable's entropy function is given by (.).

Based on the feature vectors in a class's redundancy parameters, the entropy (v) of a feature set is calculated.

$$v(f_i) = -\sum_{i=1}^{k} P(f_i) \log_2 (P(f_i)) \tag{7.2}$$

P(.) indicates the probability of the feature entering the cluster. The distance between the two observed feature vectors' entropy values determines which feature vector should be chosen. When the distance is the shortest, optimization is seen. Information gain is defined as follows:

$$\text{IGF}(f,f') = -\sum_{i=1}^{k} P(f_i) \log_2 (P(f_i)) - \left(-\sum_{i=1}^{k} P(f_i) - \sum_{i=1}^{k} P(f_i) \log_2 (P(f_i,f_i')) \right) \tag{7.3}$$

The work that is being presented eliminates low information gain features and solely keeps high information gain features. The decision is predicated on a higher IGF value than the feature value's entropy. Redundant feature vectors are eliminated during feature selection. Weighted feature selection is also utilized in feature prioritization. However, the design work enhances estimating precision and removes the functionality connected to the redundancy factor. While reducing processing overhead, low entropy feature elimination ignores the significance of feature sets. Redundancy in vector size is seen, but the variety of feature vectors in feature fusion is ignored. As a result of the similarity of the size values, this ignores crucial functionality. To overcome the problem mentioned above, a novel regression model with a divergence coefficient is suggested. The suggested strategy is demonstrated in the following section.

7.4 CLASSIFICATION FOR WEIGHTED FEATURE FUSION

Features that are short in size or extremely redundant are removed using an entropy-based feature fusion method. The significance of such functions has not, however, been linked to the significance of cardiac disease. Removal of features with high entropy and high importance can lower estimated performance; these features should be chosen depending on their significance. A feature vector divergence and its importance-based method based on its importance is a weighted feature fusion model (WFM). An autoregressive model is used to create the suggested feature selection. The autoregressive approach offers a crucial method for choosing the best combination of recorded data and ECG functionality at the same time. A data-theoretic approach was the foundation for the development of the autoregressive method. Burg's approach is employed to accomplish optimization. Based on the approximate representation of the test features by the feature vector and the registration group, this approach minimizes information loss (Gi). A feature vector set $\{f, f^{\backprime}\}$ is generated by combining the ECG signal (f_{ECG}) with the recorded physiological data value (f_R). The feature "f" is used to distinguish f′ from other features.

$$f = \{f_{ECG}, f_R\} \qquad (7.4)$$

Using the distance metric of all "j" features in the provided group, the automatic regression of the feature set is calculated to determine the variation (γi) as follows:

$$\gamma_i = (f_i - f'_{i,j})^2 \qquad (7.5)$$

In order to monitor the significance of the feature vector, weight values are linked to various forms of cardiac disease, depending on the severity factor. Heart disease severity is classified into five categories, which are presented in Table 7.1, and each category is given a weight number to represent how serious it is. According to each class's significance, weight values are allocated. Heart disease in Class 4 receives a high score of 0.5 and a healthy class of 0.1

To select the fusion function, develop a regression model using the Bergman divergence approach and select the most appropriate function value. The mathematical formula of a minimization function for feature selection optimization is as follows:

$$Arg\,min(P[\gamma_i(f,\ f')]) \qquad (7.6)$$

TABLE 7.1

Weight Association Based on the Significant Level of the Disease

Group (G_i)	Severity Category Class	Infection in %	Categorize	Weight Allotted (ω_i)
G_0	Healthy	Nil	Healthy	0.1
G_1	Category – 0	0–20%	starting level	0.2
G_2	Category – 1	20–40%	Effective with-Low	0.3
G_3	Category – 3	40–60%	Effective with –High	0.4
G_4	Category – 4	60–max %	Significant level	0.5

After the least diverse elements are eliminated, high diversity features are considered for selection. In order to determine diversity (γi) based on the square of the distance, the two feature sets f and f' are analyzed. The likelihood of variety between sets of fusion functions is defined by the probability function P (.). Calculating diversity factor involves:

$$\gamma_i\left(f,f'\right) = \left(f - f'\right)^2 \tag{7.7}$$

High diversity factor features are more likely to be chosen. However, the suggested approach evaluates the importance of the trait in a selection based on the class's level. As stated in Table 7.1, weight values are applied to class severity assessments. The feature is processed to calculate the total weighting factor (Aω) for all k classes specified as:

$$A\omega = \sum_{i=1}^{k} \omega_i \tag{7.8}$$

Where ω_i stand for the corresponding weight values for each class. The aggregated weight value for each group (c) is represented by ω_{kc}, which is calculated as follows:

$$\omega_{kc} = \sum_{i=1}^{k} \omega_{ic} \tag{7.9}$$

The choice of feature vectors for fusion is maximized by lowering the divergence factor and employing feature vectors with relatively high weight values. Optimization of feature selection is provided by:

$$Fsel \Rightarrow \left\{ \arg\min(P[\gamma_i\left(f,\ f'\right)]), \left(\left(\sum_{i=1}^{k} \omega_{ic}\right) > \frac{A\omega}{k}\right) \right\} \tag{7.10}$$

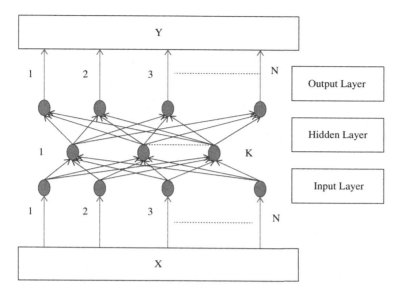

FIGURE 7.3 Basic neural network structure.

The proposed feature selection method maximizes attributes that are assigned greater weights than the average relative weights of all classes $A_{\omega/k}$. It is greatly diversified. This circumstance chooses the more significant features and denotes a change in those characteristics.

The features that are chosen will differ and have more variance and weight attached to what is sent to the classifier. The ensemble deep learning model is the basis of the classification process. During the training and testing phases, neural networks are created. The input layer, hidden layer, and output layer were all used in the construction of the neural network (NN) model, which has been trained to reduce class errors. Figure 7.3 depicts the NN model's network topology.

The chosen fusion feature xi and the weight value υi are taken into account as the NN model's input.

For, input $X = [x1, x2, ..., Xv]; \upsilon = [\upsilon1, \upsilon2, ..., Yv]$ is associated weight set. Network output comes as:

$$Y = f\left(\upsilon_t\ x\right)$$

or

$$Y = \sum_{i=1}^{v} \upsilon_i\ x_i \tag{7.11}$$

The network's output is a bipolar representation of the NN model's output.

$$Y = \begin{cases} 1,\ x \geq 0 \\ -1,\ x < 0 \end{cases} \tag{7.12}$$

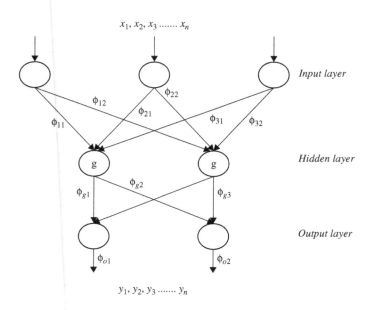

FIGURE 7.4 Structure of multilayer neural network.

Figure 7.4 shows a multi-level neural network model.

The multi-step model technique for error minimization builds a feed-forward back propagation network and is based on weight updates.

The weighted I/O relationship with the bias (g) provided to the activation function maps the output based on updates to the input (a). The relation between input and output is as follows:

$$Y_i = f\left(\sum_{i=1}^{n} \omega_i a \left(\sum_{i=1}^{k} \Phi_i x_i + g_i \right) \right) \tag{7.13}$$

The output of the NN model is a function that changes the weight variable, and the bias value after the weight update of the class is applied (g).

7.5 RESULTS

The existence or absence of cardiac disease is examined using the proposed WFM technique. An updated Cleveland database that has around 600 patient records and 14 records for each patient served as the basis for the simulation, which was created for electronic records (ER). The record was picked for fusion and updated for time-line observation. The MIT database contains the ECG signals from 600 patients. For the purpose of fusing ECG data with ER data, the ECG function is generated for peaks and periods. For the purpose of performance assessment, the suggested WFM approach is contrasted with currently used embedded fusion model approaches [27]

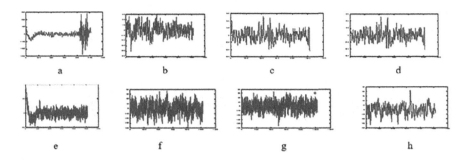

a   b   c   d

e   f   g   h

FIGURE 7.5   ECG signal inputs used from the MIT dataset.

and classifier models. Figure 7.5 and Table 7.2 display a few sets of ECG signals and ERs utilized for network training.

Section 7.4 describes a process where the properties of each ECG signal are analyzed, and a fusion strategy is selectively applied. ECG signals are randomly selected from the database for testing, and the corresponding electronic health records are accessed. Before transferring features to the classifier model, they are processed for fusion and selection. The neural network classifier is then used to classify the data and aid in disease diagnosis. Figure 7.6 displays the ECG signal after processing.

The ECG signal is analyzed to identify peaks and determine their duration. Calculations are made to determine the QRST measurement's magnitude and time window. Figure 7.6 shows a magnitude plot of the peak value.

TABLE 7.2
Fourteen Different Attributes Used as Input from Medical Records

TABLE 7.2a

ER Values	Age	Sex	Chest Pain	Trestbps	Chol	Fbs	Respect
				ER-Filed			
P1	58	0	3	144	312	1	0
P2	50	0	2	142	200	0	2
P3	64	0	3	140	313	0	0
P4	43	1	4	110	309	0	0
P5	45	1	4	128	259	0	2

TABLE 7.2b

ER Values	Thalach	Exchange	Oldpeak	Slope	Ca	Thal	Num
				ER-Filed			
P1	152	1	0	1	2	3	4
P2	134	0	0.9	1	0	7	1
P3	133	0	0.2	1	0	7	0
P4	161	0	0	2	0	7	3
P5	143	0	3	2	3	3	1

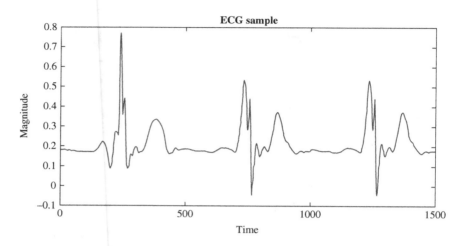

FIGURE 7.6 ECG signal input after denoising.

The average is computed, the calculated features of the processed ECG signal are listed, and variances in the computed features of the individual ECG signals are noted. The distinguishing characteristics of the various test samples are represented in Table 7.2a and 7.2b by the quantity of R peaks in the signal (F1), Q length (T1), R-R time interval duration (T2), R peak amplitude (F2), Q's position (F3), and S peak (F4) rise. P's location (F5), R's location (F6), S's location (F7), T's location (F8), and ST deviation are calculated (F9).

The WFM technique is used to process the features for ER selection and fusion. The chosen function that is supplied to the classifier model determines the classification of heart disease. To assess the effectiveness of the created technique, the system's search accuracy, search rate, accuracy, and F-score are computed for five randomly selected test iterations. These metrics are used to assess the system's accuracy.

$$\text{Accuracy} = \frac{\text{TP} + \text{TN}}{\text{TP} + \text{TN} + \text{FP} + \text{FN}} \tag{7.14}$$

The confusion matrix provided in Table 7.3 is used to obtain the four observation parameters (TP, TN, FP, and FN) as follows:

In this case,

TP is a real plus.

TABLE 7.3
Confusion Matrix

Diagnosis	Effective	Not-Effective
Effective	TP	FN
Not-Effective	TN	FP

TABLE 7.4
Result Analysis

Methods	Accuracy (%)	Recall Rate (%)	Precision (%)	F-Score (%)
SVM	84.4	81.5	87.5	84.5
L-regression	92.2	95.2	89.2	92.2
Decision tree	77.6	77.7	84.6	77.6
Naïve Bayes	83.4	78.5	88.8	83.4
Fusion model	98.5	96.4	98.2	97.2
Proposed WFM	99.2	98.3	98.9	98.4

TN is a genuine drawback.

False positives (FP) and false negatives (FN) both exist.

The following criteria describe the created system's accuracy:

$$P = \frac{TP}{TP + FP} \qquad (7.15)$$

The recall factor for the system is given by,

$$R = \frac{TP}{TP + FN} \qquad (7.16)$$

and the F-score of the system is computed given by,

$$F = \frac{2 * R.P}{R + P} \qquad (7.17)$$

The observation results of the developed system are shown in Table 7.4. The divergence and weighted features allow accuracy to be observed based on the selected features, as important observation features are retained instead of entropy measurements. The suggested system's precision has increased.

Analysis for the developed classification WFM approach is shown in Figure 7.7. The proposed method shows an accuracy improvement of 14.8% to SVM, 7% to L-regression, 21.6% to decision tree, 15.8% to Naïve Bayes, and 0.7% to fusion model. Recall rate is improved by 16.8% to SVM, 2.8% to L-regression, 20.6% to decision tree, 19.8% to Naïve Bayes, and 1.9% to fusion model. Precision is improved by 11.4% to SVM, 9.7% to L-regression, 14.3% to decision tree, 10.1% to Naïve Bayes, and 0.7% to fusion model. An F-score is improved by 13.9% to SVM, 6.2% to L-regression, 20.8% to decision tree, 15% to Naïve Bayes, and 1.2% to fusion model.

7.6 CONCLUSION

A feature fusion method based on feature vector and class attribute selection is presented. The method utilizes weighted clustering with updated features and information gain. Based on the collected observations, this method demonstrated an

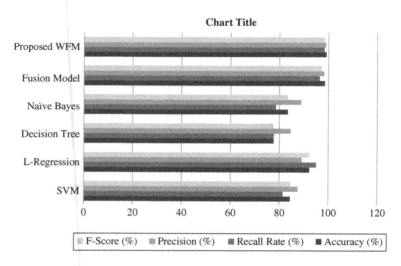

FIGURE 7.7   Analysis of classification of WFM with respective existing methods.

improvement in feature selection and an increase in the system's accuracy. Based on the chosen feature and the fusion of numerous parameters during observation, the system accuracy is increased. The important feature selection example has demonstrated the benefits of feature selection and classification effectiveness for the early detection of heart disease. Because the criticality factor is monitored based on the related weight factor, which improves system performance, the chosen feature selection strategy based on divergence and a weighted approach provides a more accurate selection of fusion features. Future iterations of the suggested method will include a wider range of cardiac monitoring scenarios, including patient situations involving multiple diseases, and a wider range of signal input conditions, including magnitude changes and signal wandering aberrations.

REFERENCES

1. Cochet H, Dubois R, Sacher F, Derval N, Sermesant M, Hocini M, Montaudon M, Haïssaguerre M, Laurent F and Jaïs P. Cardiac arrythmias: Multimodal assessment integrating body surface ECG mapping into cardiac imaging. Radiology 2014; 271:239. https://doi.org/10.1148/radiol.13131331.
2. Mandal S, Roy AH and Mondal P, Detection of cardiac arrhythmia based on feature fusion and machine learning algorithms, 2021 International Conference on Intelligent Technologies (CONIT) 2021; https://doi.org/10.1109/CONIT51480.2021.9498352.
3. Park J, An J, Kim J, Jung S, Gil Y, Jang Y, Lee K and Oh IY. Study on the use of standard 12-lead ECG data for rhythm-type ECG classification problems. Computer Methods Programs Biomed 2022; 214:106521. https://doi.org/10.1016/j.cmpb.2021.106521.
4. Lyon A, Mincholé A, Martínez JP, Laguna P and Rodriguez B. Computational techniques for ECG analysis and interpretation in light of their contribution to medical advances. J R Soc Interface 2018; 15:20170821. https://doi.org/10.1098/rsif.2017.0821.
5. Hossain AI, Sikder S, Das A and Dey A. Applying machine learning classifiers on ECG dataset for predicting heart disease. International Conference on Automation, Control and Mechatronics for Industry 4.0 (ACMI), IEEE 2021.

6. Skala T, Tudos Z, Homola M, Moravec O, Kocher M, Cerna M, Ctvrtlik F, Odstrcil F, Langova K, Klementova O and Taborsky M. The impact of ECG synchronization during acquisition of left-atrium computed tomography model on radiation dose and arrhythmia recurrence rate after catheter ablation of atrial fibrillation – A prospective, randomized study. Bratisl Lek Listy 2019; 120:177. https://doi.org/10.4149/BLL_2019_033.

7. Diwakar M, Tripathi A, Joshi K, Memoria M and Singh P. Latest trends on heart disease prediction using machine learning and image fusion. Materials Today 2021; 3213.

8. Xu Y, Luo M, Li T and Song G. ECG signal denoising and baseline wander correction based on CEEMDAN and wavelet threshold. Sensors 2017; 2754.

9. Adam AM, El-Desouky BS and Farouk RM. Modified Weibull distribution for biomedical signals denoising. Neuroscience Informatics 2022; 2, no. 1:100038.

10. Liang F, Xie W and Yu Y. Beating heart motion accurate prediction method based on interactive multiple model: An information fusion approach. BioMed Research International 2017; 1279486. https://doi.org/10.1155/2017/1279486.

11. Saeed JN and Ameen SY. Smart healthcare for ECG telemonitoring system. Journal of Soft Computing and Data Mining 2021; 2, no. 2:75.

12. Yildirim O, Talo M, Ciaccio EJ, Tan RS and Acharya UR. Accurate deep neural network model to detect cardiac arrhythmia on more than 10,000 individual subject ECG records. Comput Methods Programs Biomed 2020; 197:105740. https://doi.org/10.1016/j.cmpb.2020.105740.

13. Ullah A, Anwar SM, Bilal M and Mehmood RM. Classification of arrhythmia by using deep learning with 2-D ECG spectral image representation. Remote Sensing 2020; 12:1685. https://doi.org/10.3390/rs12101685.

14. Rajan DP, Baswaraj D, Velliangiri S and Karthikeyan P. Next generations data science application and its platform. In 2020 International Conference on Smart Electronics and Communication (ICOSEC), pp. 891–897. IEEE, 2020.

15. Solbiati M, Trombetta L, Sacco RM, Erba L, Bozzano V and Costantino G et al. A systematic review of noninvasive electrocardiogram monitoring devices for the evaluation of suspected cardiovascular syncope. Journal of Medical Devices, Transactions of the ASME 2019 Jun 1; 13:024001–1. https://doi.org/10.1115/1.4042795.

16. Almustafa KM. Prediction of heart disease and classifiers' sensitivity analysis. BMC Bioinformatics 2020; 21:278. https://doi.org/10.1186/s12859-020-03626-y.

17. Sujith AVLN, Sajja GS, Mahalakshmi V, Nuhmani S and Prasanalakshmi B. Systematic review of smart health monitoring using deep learning and artificial intelligence. Neuroscience Informatics 2022; 2, no. 3:100028.

18. Sardar P, Abbott JD, Kundu A, Aronow HD, Granada JF and Giri J. Impact of artificial intelligence on interventional cardiology: From decision-making aid to advanced interventional procedure assistance. JACC: Cardiovascular Interventions 2019; 12, no. 14:1293–1303.

19. Patil BH Dr. and Patil PM. Crow search algorithm with discrete wavelet transform to aid Mumford Shah inpainting model. Springer Journal Evolutionary Intelligence 2018; 11:73. https://doi.org/10.1007/s12065-018-0160-6.

20. Zhao W and Sampalli S. Sensing and signal processing in smart healthcare. Electronics 2020; 9:1954. https://doi.org/10.3390/electronics9111954.

21. Fontaine GH, Li G, Saguner AM and Frank R. Mechanisms of torsade de pointes tachycardia in patients with spontaneous high-degree atrioventricular block: A modern look at old data. Journal of Electrocardiology 2019; 56:55. https://doi.org/10.1016/j.jelectrocard.2019.05.007.

22. Joshi A and Shah M. Coronary artery disease prediction techniques: A survey. Lecture Notes in Networks and Systems 2020; 203:593.

23. Padeletti M, Bagliani G, De Ponti R, Leonelli FM and Locati ET. Surface electro-cardiogram recording: Baseline 12-lead and ambulatory electrocardiogram monitoring. Cardiac Electrophysiology Clinics 2019; 11:189. https://doi.org/10.1016/j.ccep.2019.01.004.
24. Ayon SI, Islam MM and Hossain MR. Coronary artery heart disease prediction: A comparative study of computational intelligence techniques. IETE Journal of Research 2020; 67:1.
25. Korra S and Sudarshan E. Smart healthcare monitoring system using Raspberry Pi on IoT platform. ARPN Journal of Engineering and Applied Sciences 2019; 14:872.
26. Trupti Vasantrao B and Rangasamy S, Weighted clustering for deep learning approach in heart disease diagnosis. International Journal of Advanced Computer Science and Applications 2021; 12, no.10:383–394.
27. Ali F, El-Sappagh S, Riazul Islam SM, Kwak D, Ali A, Imran M and Kwak K-S. A smart healthcare monitoring system for heart disease prediction based on ensemble deep learning and feature fusion. Information Fusion 2020; 63:208. https://doi.org/10.1016/j.inffus.2020.06.008.
28. Sivakumar C, Sathyanarayanan D, Karthikeyan P and Velliangiri S. An improvised method for anomaly detection in social media using deep learning. In 2022 International Conference on Electronics and Renewable Systems (ICEARS), pp. 1196–1200. IEEE, March 2022.

8 Deep Learning and Its Applications in Healthcare

Gayathiri P

CONTENTS

DOI: 10.1201/9781003345411-8

8.1 INTRODUCTION

Deep learning algorithms can be used to detect image features and diagnose diseases with greater accuracy. There exist numerous cutting-edge algorithms specifically designed for this purpose. Deep learning algorithms have been used in health and medical applications, because computed tomography (CT) or magnetic resonance imaging (MRI) in the medical profession require visual processing work. For instance, deep convolutional neural networks have been widely used in the medical field for malignant nodule categorization of lung CT. In the past, classic image processing techniques were used to reduce noise, improve image quality, create handcrafted features, or test out traditional machine learning algorithms like SVM or K-means. Deep learning techniques have greatly enhanced deep learning's capabilities recently, and it has become more and more important in the medical and health fields. Giving the computer the ability to process and analyze visual content, such as 2D, 3D, and video, is the goal of deep learning. the computer is widely used in a variety of industries, such as oil and gas [1–3], fishing and agriculture [4], image analysis in medicine [5–9], robotic surgery [10–11], and others.

Image processing can be divided into acquisition, segmentation, feature extraction, and classification [12, 13]. The most frequent deep learning issue is image categorization. Recently, machine learning-supervised learning algorithms have been used in healthcare applications [14–15], medical image segmentation, image classification [16–19], and medical image identification. We use deep learning algorithms to overcome problems for cancer detection, MRI segmentation, X-ray analysis, and COVID prediction. This chapter focuses on deep learning in healthcare applications, challenges, pros and cons, and future trends. The latest developments in deep learning- and convolutional neural network-based methods are used in medical research.

8.2 CLASSIFICATION

Image classification is a fundamental task in deep learning that has become more complete in computer-aided diagnostics used for medical image analysis. Traditionally, image classification is difficult when categorizing or labeling an image (i.e., normal case) [20, 21]. Skin cancer classification in dermoscopic images [22, 23], lung cancer detection in CT images [24], breast cancer detection in mammograms [25] and ultrasound images [26], brain cancer classification in MRI images [27, 28], diabetic retinopathy [29, 30], and eye disease recognition in retinal fundus images [31, 32] are a few clinical applications of deep learning-based medical image classification algorithms. Furthermore, the categorization of pathological images is frequently used to identify a range of malignancies, such as prostate, colon, and ovarian cancer. Deep neural networks could reach 96% sensitivity even with the small dataset for this application—just 150 cases. Deep learning methods, such as convolutional neural networks (CNN), have been used to identify COVID infections in chest X-ray images with high accuracy. CNN is a state-of-the-art classification method for medical images, and ongoing developments in DL models continue to improve its performance. Compared to existing models of medical field algorithms, the new models of algorithms give high-performance accuracy [33] and fine-tuned ResNet-50 architecture to classify skin lesions using dermatoscopic images. Some models' classification accuracy may be greatly affected by healthcare applications.

Linear discriminant analysis and an optimum deep neural network are used in [34] to classify lung cancer in CT images. It is proposed to use the conditional generative adversarial network with a simple CNN to detect breast cancer subtypes in mammograms. This network achieves 72% accuracy in classifying tumor morphology. An improved ResNet is used in [35] to classify brain cancer in MRI images. Ultra–wide-field fundus images were used to classify diabetic retinopathy by using Inception-ResNet-v2, an active deep learning architecture for image classification, to identify retinal exudates and drusen. A multiscale decision aggregation, pretrained Inception-v3, and a hybrid evolutionary deep learning model are used to classify prostate, breast, and ovarian cancer. The classification of thermography is used in medicine for various purposes, including mass screening for fever, vascular anomaly identification, and some cancer diagnoses. Thermography is a less invasive, relatively portable method of image capture to serve in diagnosis.

Thermography is a quick, noninvasive, and transportable method of gathering images for diagnostic reasons. The use of thermal cameras to spot fever in persons in public places has recently been detected, and fever screening was proposed as an early precautionary measure. Infrared technology, used efficiently during the SARS outbreak as a contact-free method of mass screening for fever detection, was used to identify 305 febrile patients out of 72,327 outpatients at a busy hospital. Deep learning algorithms have been applied to improve the accuracy of temperature readings. The classification of fever from these sensors as thermal imaging has spread into public locations more recently. In order to localize these areas, deep learning algorithms have been created. It has been found that the maximum temperature measured from the inner canthi rather than the conventional forehead scan is a stronger indication of fever.

8.3 DEEP LEARNING ALGORITHMS

Deep learning addresses the problem of understanding the content of still images or video streams to make decisions and take certain actions. To train neural networks, which are incredibly complex mathematical algorithms, deep learning datasets, or collections of images, are required. These datasets are used to teach the network how to identify patterns and features that are important for specific tasks, such as object recognition or disease diagnosis. Deep learning algorithms are being created and improved because the quantity of digital images that can be used to train neural networks on the internet is growing quickly. Major performance improvements have also been achieved. Lightning-quick CPUs drive modern machine vision systems with optimized instruction sets, GPUs with dozens or even hundreds of concurrent pipelines, and even specialized VPUs that can speed up the hardware execution of artificial intelligence (AI) algorithms. This chapter discusses deep learning in healthcare and provides a rundown of deep learning applications in the field. We will examine how end-to-end systems and new applications are being developed in important fields of medicine due to deep learning. Distributed sensors and cameras produce vast amounts of healthcare and medicine data. The potential of deep learning applications has dramatically grown with the availability of data from medical devices and digital record systems.

8.4 DEEP LEARNING IN HEALTH CARE

There are a lot of issues with fine-grained image categorization using deep convolutional neural networks and vision transformers, a system for automatically classifying skin lesions. Because skin lesions feature fine-grained heterogeneity as a distinctive feature, a trained computer can correctly identify disease by pixels. Deep learning has advanced significantly in this field, and it tries to assist computers in understanding visual input. In the past, learning algorithms that could find patterns in data required the development of feature extractors and domain knowledge.

On the other hand, deep learning is a type of representation learning consisting of multiple, sequentially organized layers. Deep learning algorithms can train highly complicated functions with high accuracy for image identification. A neural network trains to classify the image in tasks involving image classification, as shown in Figure 8.1. When a CNN is originally trained on a huge dataset unrelated to the task at hand and subsequently fine-tuned on a much smaller dataset, convolutional neural networks are shown to perform well in transfer learning. The Australian artificial intelligence mobile app Scanoma can check for the development of cancerous tumors in skin blemishes. In the meantime, dermatologists and MIT researchers created a method to identify suspicious pigmented lesions with over 90% sensitivity. A dermatologist-level cancer classification was developed using a CNN and wide-field smartphone images.

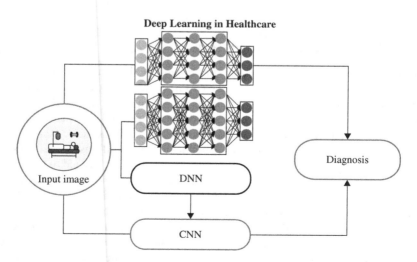

FIGURE 8.1   Deep learning in healthcare.

8.4.1 ABNORMALITY DETECTION IN RADIOLOGY

The use of deep learning in the medical industry has gained new momentum in light of the COVID-19 pandemic. This technology has shown great promise in addressing the challenges posed by the outbreak and improving healthcare outcomes. For instance, the lungs are frequently photographed. As a result, there is a noticeable increase in the need to lighten the burden on doctors and speed up the processing of X-ray images. Here, deep learning is evolving into a digital medical assistant. By analyzing medical pictures like X-rays, MRIs, and ultrasounds, Computer vision (CV) can increase the precision with which diseases are diagnosed. The CheXNet, a 121-layer convolutional neural network developed by the Stanford ML Group, was introduced for identifying pneumonia from chest X-rays. Results are more accurate than those of working radiologists. Microsoft InnerEye is another program that has significantly advanced in accurately detecting cancers on MRI images. In order to make patient treatment more accessible and accurate, Imagen and IBM Watson Imaging Clinical Review collaborated to increase physicians' ability to interpret wrist bone fractures using the potential of the largest dataset of medical pictures. Due to the deep learning application used during screening, which speeds up diagnosis and cuts down on the time patients spend having CT scans, up to 10% less radiation is generated. The power of neural network methods to eliminate extra noise and distortion from X-rays and CT images is impressive. By reducing the radiation exposure from 25% to 10%, patients are spared from having to undergo lengthy CT scans. As AI-powered computers continue to advance, they will eventually completely replace X-ray and CT machines, providing immeasurable health benefits. In non-radiation screening, there has been a substantial shift with quick outcomes.

8.4.2 DERMOSCOPIC MEDICAL IMAGE SEGMENTATION

One of the most useful techniques in medical imaging is image segmentation since it extracts a region of interest. The image is divided into sections based on the given description, which is then followed. In medical applications, segmenting body organs or tissues for border detection or to detect or segment malignancies could be a starting point. Doctors will therefore receive a result in a computer-assisted separation of the relevant area (potential disease) and background. Medical photographs often contain images with fuzzy borders, which can be difficult to recognize. To address this issue, fuzzy set theories and neutrosophic sets are used for segmentation during uncertainty handling. These techniques help to better recognize images with fuzzy borders, which are common in many types of medical imaging. Additionally, medical photographs typically consist of three colors: gray, white, and black. Look at Damae-Medical; they provide an innovative melanoma diagnosis system that aims to see beyond outward appearance.

8.4.3 DEEP LEARNING IN SURGICAL ROBOTICS

Surgical robotics, which will soon be used in many operating rooms, should not "function" without deep learning. In 2018, there were almost 900,000 robotic surgeries in the United States. First, we're talking about surgical procedures that must be performed through minuscule incisions that are challenging to produce while using human hands to grasp and control devices. Deep learning that has been trained can react to changes in the environment as quickly as a human would, acting like a human would during a task. Even if there is much precise work, typical robotic surgery still assumes that a traditional human surgeon controls robotic hands. Another instance supports the surgeon's ability to process the optimal solution to predict the tumor. For instance, during preoperative mapping, if MRI scans are taken (a series of 3D images of the brain used to characterize active parts of the brain), surgeons can see that the tumor does not cross speech and motor centers in the human brain. Here, computer and machine vision save time and lives. By combining improved deep learning, machine learning, and AI, the Boston-based startup Active Surgical is leading the charge in decreasing surgical complications related to blood loss and perfusion in real time. A growing number of applications that could potentially save patients' lives are now being developed with the help of deep learning in this field. More doctors are using this AI-powered technology to help them diagnose their patients more accurately, prescribe the best courses of action, and track the development of various diseases. Medical practitioners can save significant time by using deep learning, and patients' lives can also be saved. Technology applications for medical usage must function by expanding on how it is being used while also adding a layer of creativity and imagination. Medical imaging analysis, predictive analysis, and health monitoring, among other applications, are just a few of the areas in healthcare where deep learning is currently being used and helping medical practitioners to identify patients more accurately. You can find some advantages of deep learning technology for health systems below. Within the last six years, the popularity of the machine learning discipline of deep learning has skyrocketed. The deep learning movement is being driven by improvements in computer capacity (GPUs, parallelized computing) and the accessibility of enormous new datasets.

8.5   APPLICATIONS OF DEEP LEARNING IN HEALTHCARE

Due to the deep learning model's computational capabilities, healthcare processes are now quick, precise, and effective. Deep learning networks have a fundamental role for health systems in medical care and are revolutionizing patient care. The most frequently applied deep learning methods in the healthcare industry are CV, natural language processing, and reinforcement learning.

AI and machine learning are increasing now because of the COVID-19 pandemic. The deployment of disruptive technologies across several industries during the crisis accompanied a rapid digital revolution. The introduction of disruptive technology has the potential to improve numerous industries, including healthcare. The use of deep learning, machine learning, and AI has become critical in this field. The impact of deep learning in the healthcare industry is significant, and it has made it possible to enhance patient monitoring and diagnosis. The most innovative deep learning uses in the field of medicine are listed in Figure 8.2.

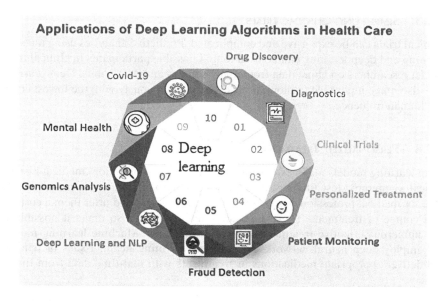

FIGURE 8.2   Applications of deep learning in healthcare.

8.5.1 DRUG DISCOVERY

Deep learning plays a big part in identifying drug formulations. Innovative tech-nologies like AI, machine learning, and deep learning supported the development of drugs and vaccines during the pandemic. Deep learning can speed up the drug discovery process, reduce costs, and simplify the difficult task of identifying poten-tial drug candidates. By analyzing large amounts of data from various sources, deep learning models can identify promising compounds, predict their efficacy, and opti-mize their properties, saving time and resources compared to traditional methods. Deep learning algorithms can predict drug characteristics, drug-target interactions, and the production of compounds with desired properties. Deep learning algorithms may easily process genomic, clinical, and population data, and several toolkits can be utilized to find patterns in the data. Researchers can now define protein structures more quickly using molecular modeling and predictive analytics thanks to machine learning and deep learning.

8.5.2 MEDICAL IMAGING AND DIAGNOSTICS

Deep learning models can interpret X-rays, MRIs, CTs, and other types of medi-cal images to make a diagnosis. The algorithms can detect anomalies in medical images, which can identify any risk. In order to identify cancer, deep learning is commonly used. Machine learning and deep learning made it possible to develop CV. It is simpler to cure various diseases with a quicker diagnosis provided by medical imaging.

8.5.3 SIMPLIFYING CLINICAL TRIALS

Clinical trials can be expensive and complicated. Predictive analytics using machine learning and deep learning can be used to find possible participants in clinical trials and let researchers combine data from many sources and data points. Deep learning will also make it possible to monitor these trials continuously with the lowest errors and human influence.

8.5.4 PERSONALIZED TREATMENT

Deep learning models make examining patient health information, medical history, critical symptoms, test findings, and other data simpler. Thus, this makes it possible for medical professionals to comprehend each patient and offer them a customized course of treatment. These innovative technologies also make it possible to find numerous effective treatments for various patients. Machine learning models can employ deep neural networks to forecast upcoming health issues or dangers and deliver appropriate medications or treatments with real-time data from linked devices.

8.5.5 IMPROVED HEALTH RECORDS AND PATIENT MONITORING

Both organized and unstructured medical and healthcare data can be processed and analyzed by deep learning and machine learning algorithms. It can be challenging to classify documents manually and keep health data accurate. Smart health records can be kept up-to-date with the help of machine learning and its subset, deep learning. There is a wealth of real-time data on health thanks to the development of telemedicine, wearables, and remote patient monitoring, and deep learning can assist in intelligently monitoring patients and identifying risks.

8.5.6 HEALTH INSURANCE AND FRAUD DETECTION

Deep learning is effective at detecting insurance fraud and predicting risks in the future. The capability of deep learning models to accurately predict patterns and behavior gives health insurance carriers another advantage when recommending comprehensive insurance plans to their clients.

8.5.7 DEEP LEARNING AND NLP

Deep learning methods are used in natural language processing (NLP) to classify and recognize words. These two technologies can recognize and categorize health data and create chatbots and speech bots. Chatbots are currently essential in telehealth. They facilitates quicker and simpler communication with patients. Chatbots were also used during the COVID-19 pandemic to respond to important queries related to the disease.

8.5.8 GENOMIC ANALYSIS

Deep learning algorithms improve biological data's interpretability and understanding. Deep learning models' advanced analytical features help researchers understand how to interpret genetic variation and design genome-based treatments. Scientists can extract properties from fixed-size DNA sequence windows using CNNs, which are often used.

8.5.9 MENTAL HEALTH

Researchers are using deep learning models to enhance clinical practice in mental health. For instance, scholarly research is still being done to comprehend how deep neural networks can be used to study the consequences of mental illness and other brain illnesses. According to researchers, trained deep learning models can outperform traditional machine learning models in specific situations. Deep learning algorithms, for instance, can learn to identify significant brain biomarkers. The aim is to develop a clinical decision support system for mental health that is affordable, digitally accessible, data-driven, and capable of implementing machine learning techniques to enhance diagnostic accuracy and patient outcomes.

8.5.10 COVID-19

The global COVID-19 pandemic has increased the need to use deep learning models. Deep learning applications are being researched to find individuals at high risk for COVID-19 and assess chest X-ray (CXR) and chest CT images to forecast intensive care unit admission and help estimate the necessity for mechanical ventilation.

8.6 ADVANTAGES OF DEEP LEARNING IN THE HEALTHCARE INDUSTRY

Deep learning can effectively assist any medical task that requires a skilled eye to identify and categorize a health issue. This AI-supported system reads images in real time and uses sophisticated algorithms to aid image processing in identifying specific indicators of sickness. The right application of deep learning in medicine will aid in minimizing the time spent on pointless diagnostic processes and give the medical practitioner the tools needed to make more precise diagnoses and prescribe more efficient treatments. Deep learning has developed to the point where it can assist with a wide range of jobs, including many medical procedures. As a result, an increasing number of specialists can now benefit from various advantages, as illustrated in Figure 8.3.

8.6.1 A MORE ACCURATE METER

Diagnoses are made more quickly and precisely when deep learning is used in healthcare applications. The system's accuracy rates increase with the amount of data it gets for algorithm training. For instance, a Santa Clara Valley medical center

A More Accountable   Early Disease   Enhanced Medical
Meter   Recognition   Procedures Efficiency

Automatic Generation   Interactive Medical
of Medical Reports   Imaging

FIGURE 8.3 Advantages of deep learning in healthcare.

found that computer-based algorithmic systems could estimate blood loss after cesarean deliveries more precisely than previous approaches. This means that algorithms based on neural networks can be trained to recognize items with higher levels of accuracy than medical professionals can and to spot patterns that the human eye might otherwise miss. X-rays, ultrasound, CT scans, and MRI are other imaging methods that show better accuracy.

8.6.2 EARLY DISEASE RECOGNITION

Many diseases only respond to medical intervention in their early stages. Deep learning technology makes it possible to identify symptoms before they become obvious and helps doctors act quickly. This significantly impacts the treatment of individuals who would not otherwise receive the assistance they require. By identifying illnesses early, medical professionals can give medications to combat those conditions or even perform procedures sooner and save lives. Deep learning is intended to hasten the diagnosis process and improve treatment efficacy.

8.6.3 ENHANCED MEDICAL PROCEDURES EFFICIENCY

In addition to being generally effective for patients and healthcare workers, deep learning is renowned for its diagnostic precision. In light of the projected physician shortage, this reduction is very advantageous. By 2032, there may be a 121,900-physician shortfall in the United States alone. The team's workflow may be streamlined by using deep learning on digital photos, ultimately leading to cost optimization. Advanced AI algorithms handle most labor-intensive tasks, freeing the specialist to concentrate on relationship development, information gathering, counseling, and management.

8.6.4 AUTOMATIC GENERATION OF MEDICAL REPORTS

The development of deep learning has made it possible to use medical imaging data extensively for more specific illness diagnosis, therapy, and prediction. By utilizing deep learning techniques, medical personnel can gather improved medical data that can be used for disease prediction and report generation in addition to diagnosis and drug prescription. Healthcare specialists can leverage the power of deep learning to automate medical report generation. By feeding data from X-rays, ultrasounds, CT scans, and MRI to deep learning algorithms, clinicians can gain in-depth insight into an individual's physical condition, predict when a disease will develop, and when appropriate treatment will be needed.

8.6.5 INTERACTIVE MEDICAL IMAGING

Medical imaging with deep learning enables interactive, in-depth 3D visualization. The use of deep learning techniques in medical image analysis has greatly benefited it in recent years. These innovations opened the path for deep learning to improve its efficiency in processing medical images. In order to provide more precise medical diagnoses, deep learning may now be combined to visually examine dynamic 3D models. Compared to conventional 2D photos, 3D models are a more educational medium. As a result, 3D breast imaging powered by cutting-edge deep learning technologies is more successful at preventing cancer in the early stages of the disease.

8.7 USES

Machine learning is a subset of deep learning. Commonly referred to as deep structured learning or hierarchical learning, it is loosely based on how neurons interact with one another in animal brains to process information. Artificial neural networks, a layered algorithmic design used in deep learning, evaluate data to mimic these connections. A deep learning algorithm can learn to find correlations and connections in the data by analyzing how data is filtered through an artificial neural network's (ANN's) layers and how the layers connect. These features make deep learning algorithms cutting-edge tools with the potential to transform healthcare. The most common types of uses are given below.

8.7.1 DEEP NEURAL NETWORKS

A deep neural network (DNN) is a kind of ANN and is differentiated from other neural networks by the number of layers that make up its depth. These layers carry out the mathematical translation operations that transform raw input into useful output. In order to further improve the output, additional layers enable slightly different translations to be executed in each layer. These algorithms can predict pediatric appointment no-shows, as demonstrated by a recent DNN use case example. Electronic health record (EHR) data and local weather data can serve as predictors for pediatric patients who may miss their appointments. By developing a prediction model that incorporates these data, physicians can implement preventive interventions to reduce the rate of no-shows.

The primary care pediatric clinic at Boston Children's Hospital provided the researchers with EHR data for 19,450 patients between January 10, 2015, and September 9, 2016, which they used to create their model. 20.3% of the 161,822 appointments corresponding to these patients were missed. On the day of the appointment, information on the local weather was also shared. The DNN performed better than the traditional no-show prediction method, showing that the weather and a patient's past no-show records were the two key indicators of whether they would appear for their upcoming appointment.

Using a DNN, researchers could more quickly assess the correlation between several aspects of a patient no-show and identify the aspects that were most closely associated with the result.

8.7.2 CONVOLUTIONAL NEURAL NETWORKS

DNNs using CNNs are used to analyze visual data. CNNs examine photos and glean attributes they can use to classify the images. In medical imaging, classification is essential when a physician will examine an image, such as a CT scan or an X-ray, to identify a range of illnesses. A CNN algorithm could help with medical imaging activities, enhancing clinical decision support and solving population health problems.

Researchers wanted to create a CNN model that could concurrently detect several retinal disorders in a study examining the link between retinal disease and lack of access to care. They created their model using 120,002 images of the eye's fundus, which were then examined and labeled according to the diagnosis of retinal diseases by a team of professional ophthalmologists. The model's performance was compared to that of a team of retinal specialists in order to assess the model's capability to identify the retinal illness.

In seven of the 10 retinal disorders under consideration, CNN's accuracy was better than or on a par with that of the expert group. When comparing image analysis speed, the model fared better than doctors. It completed one image analysis in under a second compared to the fastest specialist's 7.68 seconds.

The researchers posited that the success of their model could help address access to care in underdeveloped regions where lower screening rates and late diagnosis of retinal disease are associated with an increased risk of irreversible vision loss.

8.7.3 RECURRENT NEURAL NETWORKS

Recurrent neural networks (RNNs) are an additional ANN type that uses sequential or temporal input. They are frequently employed for speech recognition, image captioning, natural language processing, and translation. RNNs use information from inputs in earlier layers to affect the present inputs and outputs, in contrast to other neural networks where inputs and outputs are independent of one another.

For activities like choosing the cohort for clinical studies, RNNs are helpful for healthcare professionals. Clinical trials involve the selection of a cohort or group of patients for study involvement who share comparable pertinent characteristics. This can be a time-consuming and expensive process, because the outcome of a clinical study depends on the careful selection of a patient cohort.

To explore if different DL models could effectively identify crucial characteristics for cohort selection, researchers set out to test this hypothesis to save time and money. A straightforward CNN, a deep CNN, an RNN, and a hybrid model fusing both CNN and RNN were all trained and put to the test. Models were given patient records manually labeled by specialists to state whether patients met one or more requirements out of 13 possibilities for a clinical trial. Overall, the hybrid and RNN models performed much better than the CNN models. However, the researchers acknowledged that the tiny dataset they used for this study had limitations. They suggested additional research before a cohort selection model could be applied.

8.7.4 GENERATIVE ADVERSARIAL NETWORKS

Generic data that can be used in place of real data is produced by generative adversarial networks (GANs), which combine the output of two neural networks. Image, video, and voice generation frequently make use of GANs. GANs' capacity to produce fake MRI pictures makes them extremely useful in the healthcare industry. Researchers face various difficulties when using medical images to train AI models for diagnosis and predictive analytics because of the variability in their quality, the potential for patient privacy concerns, and the frequently unbalanced nature of image datasets. In order to allay these worries, researchers have looked into training deep learning models for clinical decision assistance utilizing MRI images generated by GANs. In order to create fake data that can be utilized in place of genuine data, GANs combine the output of two neural networks. In image, video, and voice generation, GANs are frequently employed.

Because they can produce fake MRI images, GANs have a lot of potential for medical use. Medical images can be of varying quality, patient privacy laws may govern them, and image databases are frequently unbalanced. These factors make it difficult for researchers to use medical images to train AI models for diagnoses and predictive analytics. Researchers have researched deep learning models for clinical decision support using MRI images generated by GANs to address these problems. Intelligent deep learning algorithms can develop the ability to recognize complex patterns through practice on cases that have already been identified. Today, deep learning is used in many medical specialties and continues to improve healthcare.

8.8 SUMMARY

The healthcare sector is leading the way in innovation and development. However, the problem of accelerating time to market for new products will never go away. Additionally, few businesses have the means to make innovations like deep learning technology available to the general population. By improving diagnoses and treatments in terms of precision, reproducibility, and scalability, deep learning-based algorithms will complement the conventional function of medical practitioners. Table 8.1 shows the merits and demerits of deep learning algorithms.

TABLE 8.1

Merits and Demerits of Deep Learning Algorithms

S. No.	Reference No.	Algorithm	Merits	Demerits
1	[31]	Back Propagation	• Flexible and effective. • Simple to implement	• When used with test data, it might not perform well. • It might be sensitive to noisy data. In local optima, it frequently becomes stuck. If the training set is too small, overfitting may occur.
2	[32]	CNN	• Memory is saved, and fewer parameters are used to create a 2D structure from an input image using local connections and weights; quick to train; able to learn relevant elements from an image or video at multiple levels. • They perform better and are particularly effective at extracting features. They are effective at providing pre-training and are simple to use practically.	• Need a huge memory requirement to keep all the intermediate results. During the early training phases, they frequently misclassify things and become puzzled by visuals. If the data were entirely inaccurate, they would be useless. • Data that is unstructured can be very difficult to process.
3	[33]	DNN	• The labeling of data can be expensive and time-consuming. • It is capable of understanding and applying unique data. A deep learning technique eliminates the need for well-labeled data, because the algorithms are excellent at learning without rules.	• Data analysis is an important step in the deep learning training process. However, there is a temporal limit on how long the training process can last because of the fast-moving and streaming input data. • Due to the major differences between each application and the wide variety of testing techniques for analysis, validation, and scalability, evaluating a system's performance in real-world settings can be challenging.

(Continued)

TABLE 8.1 (Continued)
Merits and Demerits of Deep Learning Algorithms

S. No.	Reference No.	Algorithm	Merits	Demerits
4	[34]	GAN	• The generator model does not need to define the shape of the distribution probability. • Data generated by sampling can be parallelized. There is no need to estimate a probability when reporting a lower bound, as in VAE.	• There is no distinct representation. It must be properly coordinated through training. Between the stages of learning, information should be updated quietly, like the negative chains for a Boltzmann machine. • A decrease in the cost function of the generator causes an increase in the cost function of the discriminator and vice versa. • Instability and a failure of convergence for the GAN game could have occurred. • The issue of mode collapse.
5	[35]	RBN	• They are feature extractors and can perform pattern completion. Useful for training other models. • It can be stacked to train a deeper feed forward neural model beforehand.	• Effective RBM training is challenging. Unable to monitor the required loss.
6	[36]	DBN	• Even for models with many parameters and nonlinear layers, it is possible to learn an ideal set of parameters quickly. • An approximate inference process can calculate the output values of variables in the lowest layer.	• The approximation inference procedure's single bottom-up pass restriction, the fact that it never readjusts for changes in the network's other layers or parameters, and how slowly and ineffectively it works.

REFERENCES

1. Elyan, E., Jamieson, L. and Ali Gombe, A. (2020). Deep learning for symbol detection and classification in engineering drawings. *Neural Networks, 129*, 91–102.
2. Jain, A., Gandhi, K., Ginoria, D. K. and Karthikeyan, P. (2021, December). Human activity recognition with videos using Deep Learning. In *2021 International Conference on Forensics, Analytics, Big Data, Security (FABS)* (Vol. 1, pp. 1–5). IEEE.
3. Moreno-García, C. F., Elyan and Jayne, C. (2017). Heuristics-based detection to improve text/graphics segmentation in complex engineering drawings. In *International Conference on Engineering Applications of Neural Networks* (pp. 87–98). Springer, Cham.
4. Anthoniraj, S., Karthikeyan, P. and Vivek, V. (2022). Weed detection model using the Generative Adversarial Network and Deep Convolutional Neural Network. *Journal of Mobile Multimedia, 18*(2), 275–292.
5. Schwab, E., Gooßen, A. and Deshpande, H., (2020). Localization of critical findings in chest X-ray without local annotations using multi-instance learning. In *2020 IEEE 17th International Symposium on Biomedical Imaging (ISBI)*, Iowa City, IA (pp. 1879–1882).
6. Pomponiu, V. and Nejati, (2016). Deep neural networks for skin mole lesion classification. In *2016 IEEE International Conference on Image Processing (ICIP)*, Phoenix, AZ (pp. 2623–2627).
7. Esteva, A., Kuprel, B., Novoa, R.A., Ko, J., Swetter, S.M., Blau, H.M. and Thrun, S. (2017). Dermatologist level classification of skin cancer with deep neural networks. *Nature, 542*(7639), 115–118.
8. Periyasami, K., Viswanathan Mariammal, A. X., Joseph, I. T. and Sarveshwaran, V. (2020). Combinatorial double auction based meta-scheduler for medical image analysis application in grid environment. *Recent Advances in Computer Science and Communications (Formerly: Recent Patents on Computer Science, 13*(5), 999–1007.
9. Schlemper, J., Oktay, O., Schaap, M., Heinrich, M., Kainz, B., Glocker, B. and Rueckert, D. (2019). Attention gated networks: Learning to leverage salient regions in medical images. *Medical Image Analysis, 53*, 197–207.
10. Vyborny, C. J. (1994). Can computers help radiologists read mammograms? *Radiology, 191*(2), 315–317.
11. Sevugan, A., Karthikeyan, P., Sarveshwaran, V. and Manoharan, R. (2022). Optimized navigation of mobile robots based on faster R-CNN in wireless sensor network. *International Journal of Sensors Wireless Communications and Control, 12*(6), 440–448.
12. Gonzalez, R. C. (2009). *Digital Image Processing.* Pearson Education India.
13. Szegedy, C. et al., Going deeper with convolutions. In 2015 IEEE Conference on Computer Vision and Pattern Recognition (CVPR), Boston, MA, USA (pp. 1–9). doi: 10.1109/CVPR.2015.7298594).
14. Krizhevsky, A. and Sutskever I. (2012). Imagenet classification with deep convolutional neural networks. *Advances in Neural Information Processing Systems, 25*, 84–90.
15. Velliangiri, S., Anbarasu, V., Karthikeyan, P. and Anandaraj, S. P. (2022). Intelligent personal health monitoring and guidance using long short-term memory. *Journal of Mobile Multimedia*, 349–372.
16. Deepajothi, S., Rajan, D. P., Karthikeyan, P. and Velliangiri, S. (2021, March). Intelligent traffic management for emergency vehicles using convolutional neural network. In *2021 7th International Conference on Advanced Computing and Communication Systems (ICACCS)* (Vol. 1, pp. 853–857). IEEE.
17. Gouda, W., Sama, N. U., Al-Waakid, G. and Humayun, (2022). Detection of skin cancer based on skin lesion images using deep learning. In *Healthcare* (Vol. 10, No. 7, p. 1183). Multidisciplinary Digital Publishing Institute.

18. Rajinikanth, V., Aslam, S. M. and Kadry, S. (2021). Deep learning framework to detect ischemic stroke lesion in brain MRI slices of flair modalities. *Symmetry, 13*(11), 2080.
19. Shvets, A. A., Rakhlin, A. and Kalinin, A., (2018). Automatic instrument segmentation in robot-assisted surgery using deep learning. In *2018 17th IEEE International Conference on Machine Learning and Applications (ICMLA)* (pp. 624–628). IEEE.
20. Sarvamangala, D. R. and Kulkarni, R. V. (2021). Convolutional neural networks in medical image understanding: a survey. *Evolutionary Intelligence, 15*, 1–22.
21. Sarker, M. M. K., Makhlouf, Y., Banu, S. F. and Chambon (2020). Web-based efficient dual attention networks to detect COVID-19 from x-ray images. *Electronics Letters, 56*(24), 1298–1301.
22. Yap, J. and Yolland (2018). Multimodal skin lesion classification using deep learning. *Experimental Dermatology, 27*(11), 1261–1267.
23. Carvalho, R., Morgado, A. C., Andrade, C. and Nedelcu, T. (2021). Integrating domain knowledge into deep learning for skin session risk prioritization to assist teledermatology referral. *Diagnostics, 12*(1), 36.
24. Tian, P., He, B., Mu, W., Liu, K. and Liu, Y (2021). Assessing PD-L1 expression in non-small cell lung cancer and predicting responses to immune checkpoint inhibitors using deep learning on computed tomography images. *Theranostics, 11*(5), 2098.
25. Singh, V. K., Romani, S., Rashwan, H. A., Akram, F. and Pandey, N. (2018). Conditional generative adversarial and convolutional networks for X-ray breast mass segmentation and shape classification. In *International Conference on Medical Image Computing and Computer-Assisted Intervention* (pp. 833–840). Springer, Cham.
26. Hijab, A., Rushdi, M. A. and Gomaa, M. (2019). Breast cancer classification in ultrasound images using transfer learning. In *2019 Fifth International Conference on Advances in Biomedical Engineering (ICABME)* (pp. 1–4). IEEE.
27. Ismael, S. and Mohammed, A. (2020). An enhanced deep learning approach for brain cancer MRI images classification using residual networks. *Artificial Intelligence in Medicine, 102*, 101779.
28. Hashemzehi, R., Mahdavi, S. J. S. and Kheirabadi, M. (2020). Detection of brain tumors from MRI images based on deep learning using hybrid model CNN and NADE. *Biocybernetics and Biomedical Engineering, 40*(3), 1225–1232.
29. Qureshi, I. and Ma, J. (2021). Diabetic retinopathy detection and stage classification in eye fundus images using active deep learning. *Multimedia Tools and Applications, 80*(8), 11691–11721.
30. Martinez-Murcia, F. J., Ortiz, A., Ramírez, J. and Górriz, J. (2021). Deep residual transfer learning for automatic diagnosis and grading of diabetic retinopathy. *Neurocomputing, 452*, 424–434.
31. Hasanraza, A. N. S. A. R. I. (2020). Artificial neural network: Learn about electronics (learn electronics). [(accessed on 29 May 2022)]. Available online: libgen.li/file.php?md5=a3245497addad0753ea49636a9777acc.
32. Hasan, M., Alam, M., Elahi, M., Toufick, E., Roy, S. and Wahid, S.R., 2020. CVR-Net: A deep convolutional neural network for coronavirus recognition from chest radiography images. *arXiv preprint arXiv:2007.11993*.
33. Hossain, M. S. and Muhammad, G. (2020). Deep learning-based pathology detection for smart connected healthcare. *IEEE Network, 34*(6), 120–125.
34. Al-Askar, H., Radi, N. and MacDermott, Á., 2016. Recurrent neural networks in medical data analysis and classifications. In *Applied Computing in Medicine and Health* (pp. 147–165). Morgan Kaufmann.
35. Arora, A. and Arora, A., 2022. Generative adversarial networks and synthetic patient data: current challenges and future perspectives. *Future Healthcare Journal, 9*(2), 190.
36. Ahlawat, N. and Vinod, D.F., 2022. Clipped RBM and DBN Based Mechanism for Optimal Classification of Brain Cancer. In *ICT with Intelligent Applications: Proceedings of ICTIS 2022*, Volume 1 (pp. 295–304). Singapore: Springer Nature Singapore.

9 Future of Computer Vision Application in Digital Healthcare

Rajendra Kumar Bharti
Mohiuddin Ali Khan
Rajadurai Narayanamurthy
Prarthita Biswas
Himanshu Sharma

CONTENTS

DOI: 10.1201/9781003345411-9

9.1   INTRODUCTION

Artificial intelligence and deep learning include a subfield called computer vision, where humans teach computers to observe and understand their environment. Even though people and animals naturally handle the problem of vision from a very young age, helping robots perceive and sense their surroundings through vision is still a mostly unsolved subject. Due to the limits of human vision and the infinitely variable terrain of our dynamic world, machine vision is fundamentally challenging [1, 2].

9.1.1   HOW DOES COMPUTER VISION WORK

Various steps are taken into account in order to explain the working of computer vision, and these start with the procedures below.

9.1.1.1   Acquiring an Image

We need to get, capture, and acquire images that can be captured in real time or with the help of videos and three-dimensional (3D) technology. The various images captured are required for analysis.

9.1.1.2   Processing Acquired Images

Deep learning processes the images, and the various models used are listed below.

1. Convolutional neural networks (CNNs).
2. Generative adversarial networks (GANs).
3. Long short term memory networks (LSTMs).
4. Recurrent neural networks (RNNs).
5. Radial basis function networks (RBFNs).
6. Multilayer perceptrons (MLPs).
7. Restricted Boltzmann machines (RBMs).
8. Autoencoders.
9. Self-organizing maps (SOMs).
10. Deep belief networks (DBNs).

9.1.1.3   Understanding the Image

The final step is to identify the image, which is processed so that it can be used for interpretation.

9.2   THE SIX STEPS TO CREATING A SUCCESSFUL COMPUTER VISION POC (PROOF OF CONCEPT)

The application cases for computer vision models are incredibly varied, boosting company performance, automating crucial decision-making processes, etc. But if a promising model doesn't behave as anticipated, it might be an expensive liability. A

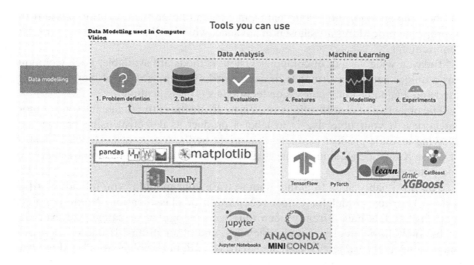

FIGURE 9.1   Six steps for creating a successful computer vision.

- Identify the business problem.
- Define the success criteria.
- Determine the appropriate computer vision techniques.
- Collect and label training and test images.
- Train and evaluate model.
- Deploy and test and iterate on the solutions

comprehensive overview of computer vision, one of the main subfields of artificial intelligence, is shown in Figure 9.1. Using computational methodologies, computer systems may analyze digital photos, movies, and other visual data with AI vision to generate knowledge from which to act or make judgments [3].

9.2.1 Identify the Business Problem

Many vision projects fall short of their goals because they focus on the incorrect issue. A vision project needs a distinct corporate aim and benefits to be effective. In one to two sentences, you should be able to summarize the objective and benefit. Reduced stock outs on shop shelves or fewer defective items leaving factories are two examples of goals that could be set. Spend time with business partners familiar with the issue if the benefits or business challenges are unclear [4].

9.2.2 Define the Success Criteria

Spend some time considering the business process and how to calculate the business benefit, and this stage ought to be simple. Here, the objective is to translate the business outcome into specific success criteria that can be used to gauge how well the solution works. Finding the strategies that best fit the problem before gathering data or choosing an algorithm. Your team can concentrate on the execution phase if the appropriate procedures are chosen in advance, which will help you define your data requirements.

Today, a wide variety of computer vision methods are available. Each technique has a unique set of performance measures and algorithms. One of the most widely used and adaptable computer vision techniques is classification.

The objective of classification issues is to divide the entire image into one or more categories. If your objective is to label a product as defective using photographs showing a single product example, this might be the tool to use. The classification of chest X-rays depending on the degree of illness damage is a good example of computer vision.

You'll probably need to gather data to train your model if you intend to utilize a deep learning model for object recognition or classification. Numerous deep learning models have already been taught to recognize or categorize everyday items, including vehicles, people, bicycles, and other things. If one of these frequent items is the subject of your scenario, you might be able to quickly download and deploy a pretrained model. If not, you'll need to gather and label data to train your model.

You are prepared to move on to model training once a quality collection of photos has been labeled. The next stage is applying transfer learning to train your computer vision model, assuming you are using a deep learning model. The neural network is frozen in transfer learning, and only the last SoftMax layer is retrained to adapt a previously trained model to a new situation. With little data, transfer learning is a quick and efficient method for training deep learning vision models. Other methods, such as fine-tuning, have higher data requirements but give a more advanced performance.

9.3   TRENDS IN COMPUTER VISION

In recent years, rapid prototyping and 3D modeling technologies have pushed the development of medical imaging modalities like CT and MRI. One of the most cutting-edge and constantly expanding fields is computer vision. Grand View Research estimates that the size of the worldwide computer vision market was $11.32 billion in 2020 and will increase at a compound yearly growth rate of 7.3% from 2022 to 2028. Facial recognition software, which enables computers to match images of people's faces to their identities, heavily relies on computer vision [5–7].

Computer vision algorithms find facial characteristics in photos and contrast them with face profile databases. As is well known, the medical industry generates enormous amounts of data from various medical equipment, digital record systems, and distributed sensors and cameras, greatly enhancing the potential of deep learning [8]. Figure 9.2 depicts how the computer vision is applied in healthcare.

As it deals with teaching the computer to understand data in the form of an image, the use of deep learning techniques will result in significant technological

TRUE/Original

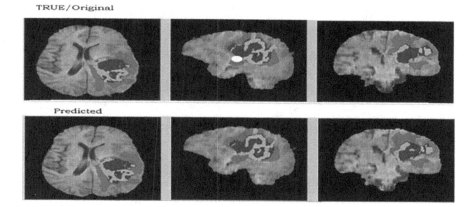

FIGURE 9.2 How computer vision can be applied in healthcare.

advancements. Deep learning will take care of finding the necessary features in graphics instead of a human doing so. Deep learning is a type of representation learning made up of numerous layers of representations ordered progressively [9–11].

Since computer vision focuses on comprehending images and extracting their characteristics, this includes comprehending videos and extracting features from them. It requires executing the duties of object identification, image classification, and segmentation [12]. Finally, it predicts whether the image is normal, as shown in Figures 9.3 and 9.4.

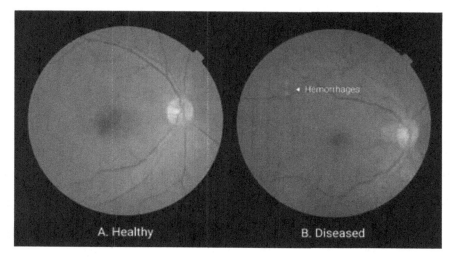

FIGURE 9.3 Healthy and diseased images.

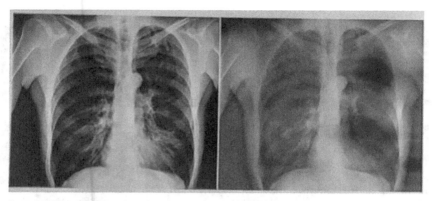

Input: Chest X-Ray Image   **Output:Pneumonia Positive**

FIGURE 9.4   Chest X-ray.

Many improvements in object detection and picture classification have been made, and these improvements will enhance medical imaging. Numerous studies showed that managing complicated medical tasks linked to diagnostics in radiology, pathology, and dermatology produced extremely accurate and precise outcomes.

9.3.1   POPULAR COMPUTER VISION TOOLS

1. TensorFlow
2. CUDA
3. MATLAB
4. Keras
5. OpenVINO
6. DeepFace
7. YOLO
8. OpenCV
9. Viso Suite
10. SimpleCV
11. BoofCV
12. CAFFE

9.3.2   MOST COMMON DATASETS USED

- The CIFAR-10 dataset and 4.
- The CIFAR-100 dataset.
- IMDB-Wiki dataset
- ImageNet
- PASCAL VOC dataset
- LabelMe dataset
- MS COCO dataset
- Places2 database

- Fashion MNIST
- MNIST database

9.3.3 COMMON TASKS ASSOCIATED WITH COMPUTER VISION

Computer vision issues instruct computers to understand digital images and visual data from the outside environment. To conclude, it could be necessary to collect, process, and analyze data from such inputs. The extensive formalization of difficult issues characterized the development of machine vision into well-known, convincing problem statements.

9.4 CLASSIFICATION OF IMAGE

One of the topics that has received the greatest attention since the launch of the ImageNet dataset in 2010 is image categorization. The most frequent computer vision task is image classification, which involves assigning a label or category to an image based on its visual features. Image classification is a common computer vision task that involves assigning a specific label to an entire image from a given collection of photos based on a predefined set of classes. Unlike more complex problems such as object recognition and image segmentation, which require the localization of detected features by assigning coordinates, image classification focuses on analyzing the whole image [11, 13, 14].

9.4.1 DETECTING OBJECTS

Object detection, as the name implies, deals with identifying and positioning things using bounding boxes. Class-specific traits are looked for in an image or video and detected by object detection whenever they appear. Depending on what the detection model was trained on, these classes may include people, animals, or vehicles.

9.4.2 SEGMENTATION OF IMAGES

An image is segmented into smaller sections called sub-objects to show that a machine can distinguish an object from its surroundings and/or another object in the same image.

9.4.3 PERSON AND FACE RECOGNITION

The human face is the main object detected by facial recognition, a subset of object detection. In that task, when features are located and detected, facial recognition is similar to object detection because it involves both the detection and recognition of the discovered face. Facial recognition algorithms classify faces based on the positioning of these landmarks and common features like the eyes, mouth, and nose. Haar Cascades, which is easily accessible through the OpenCV library, is one of the facial recognition methods based on conventional image processing methods. More dependable deep learning-based algorithmic techniques are discussed in works like FaceNet. An "image segment" refers to a specific category of objects identified by a neural network in an image, which can be extracted using a pixel mask [15, 16].

9.4.4 EDGE DETECTIONS

The task of identifying boundaries in objects is known as edge detection. An algorithm is employed to perform this work, and it uses mathematical methods that aid in detecting abrupt changes or discontinuities in the image's brightness. Convolutions with specially designed edge detection filters and traditional image processing-based techniques like Canny Edge Detection are the two main approaches to detecting edges. It is widely applied as a stage in data preprocessing for many jobs.

Additionally, edges in an image give us crucial details about the image's contents. All deep learning algorithms use edge detection by default to collect global low-level features using learnable kernels.

9.4.5 IMAGE RESTORATION

Image restoration is the process of restoring or reconstructing old, fading copies of images that were improperly captured and preserved, resulting in the loss of the image's quality.

9.4.6 FEATURES MATCHING

Features in computer vision are regions of an image that reveal the most details about a particular object in the image. Edges are strong indicators of object detail, making them essential features, even if corners and even more localized, sharp details, like edges, can also serve as features. Using feature matching, we may compare the features of similar regions in one image to those in another.

Today we make use of computer vision in various applications.

1. Monitoring of remote patients.
2. Detection of cancer in early stages.
3. Detection of tumors.
4. Machine-assisted diagnosis.
5. Timely detection of diseases.
6. Medical imaging and training
7. X-ray analysis.
8. CT and MRI analysis.

9.4.6.1 Role of Computer Vision

Computer vision has enormous potential for use in healthcare and medical imaging in the modern world. But since technology develops quickly, more and more medical use cases are now feasible. However, computer vision in healthcare applications must work with privacy-preserving deep learning and picture recognition. The significance of computer vision, a branch of artificial intelligence, will grow in the coming days, as it has proven useful for diagnosis and getting a second opinion on an issue [17, 18].

The fundamental functions of computer vision are obtaining the images needed to perform the task, processing them by extracting the data from the images, and analyzing the results. After extracting multi-dimensional data from the captured images, the data is analyzed to identify trends and patterns.

9.5 PHASES OF COMPUTER VISION

Digitization and signal processing are two steps of computer vision. Images from the actual world are collected using a video camera and then converted to digital form to be processed and cleaned using signal processing software.

Using different filters, noise problems in the signal processing process can be reduced to reduce blurring effects. The next step, considered the core of most computer vision, is edge detection, because of how these edges contribute significantly to human visual comprehension.

A step known as area detection comes next, where a region is a collection of connected pixels with essentially the same intensity. The main purpose of this step is to locate the dominant shapes in the obtained image. After extracting all the features from an image, we must determine the various objects in the last phase. This will result in some kind of semantically symbolic representation.

The medical field will benefit greatly from using computer vision in numerous applications to deliver better and more timely services as the number of patients and the time needed to manually diagnose them rises.

The use of computer vision will increase on a daily basis and become part and parcel of the doctor's daily routine in handling various diseases and diagnosing them.

Recent technological developments have made it clear that the healthcare sector will gain from them because they require less time and effort to determine the patient's health or medical condition.

With the use of computer vision, it is quite likely that computers are designed to see in order for them to observe objects, form opinions, and gather data. The ancient adage "a picture is worth a thousand words" certainly applies. The collected photographs will be broken down into pixels by computer vision, which will then assist in matching the patterns created.

9.6 ADVANTAGES OF COMPUTER VISION

One of the many benefits of using computer vision in healthcare is that the application will offer a quicker, better, and more accurate diagnosis. The more data we supply, the better the machines spot different patterns. This will make it easier for neural network algorithms to learn properly and make decisions faster and more accurately than humans [19, 20].

The second benefit that we observe is early illness detection. Computer vision technology makes it possible to identify symptoms before they become obvious and helps doctors act quickly. This significantly impacts the treatment of individuals who would not otherwise receive the assistance they require. By identifying illnesses early, medical professionals can give medications to combat those conditions or even perform procedures sooner and save lives. The use of computer vision is intended to hasten the diagnosis process and improve the efficacy of treatment [21, 22].

The other advantages include increasing the efficiency of medical procedures and automatic generation of medical reports.

More and more people will make use of this technology, as it is capable of learning on a day-to-day basis, as and when the data is provided. Also, the intelligent computer vision algorithm learns from the previous cases which are diagnosed. Treatment plans are completed successfully. Because of this, the use of this technology will never decrease. Instead, it will skyrocket as the need and demand will increase.

The healthcare sector, which is at the forefront of innovation and transformation, is well aware that it is going through a transitional period in which every component is connected to technology. Along with this, there will be difficulties in getting people to buy the newly created products and marketing them.

The development of technologies like computer vision, which makes it possible to recognize and detect images in real time due to the processing speed available, will be determined by a number of factors. Computer vision accuracy will improve as dataset quality continues to rise.

Finally, we can say that using computer vision applications will benefit both patients and doctors in terms of reducing the time needed to identify the issue, reducing diagnosis errors, and detecting minute anomalies and deviations from the norm that an expert in physical observations would otherwise miss.

In addition, it aids in the preparation needed before surgery is scheduled for the doctors, nurses, and patients. The use of technology will make it easier to monitor different surgical tools both during and after the procedure. It also aids in the training of aspiring surgeons.

The various computer vision medical applications enable speedier admissions, self-service kiosk accessibility, remote health monitoring scenarios, and other advantages of medical automation for patients. However, the most crucial point is that using computer vision in healthcare scenarios saves lives and reduces the intensity, trauma, and cost of treatments.

With computer vision, objects will likely be seen or observed with more focus and a wider range of view than by human eyes, allowing it to simultaneously see many targets. The ability of computer vision to view objects using several means, such as infrared light and X-ray imaging, makes it feasible to perceive images that are exceedingly challenging for human eyes to see.

Computer vision can also understand high-resolution and high-dimensional images containing more information than ordinary images. Second, the advantage of computer vision is reflected in its higher precision and speed when processing vision signals.

Endless state-of-the-art algorithms have gifted computer vision the power to detect and classify objects from input images, which has been demonstrated to beat human capacity in many aspects. Due to these advantages, computer vision technology has been deployed in health and medical applications.

Many situations involving CT or MRI in the medical profession require work in image processing; for instance, deep convolutional neural networks have been widely used in medical imaging processing for categorizing benign and malignant nodules in lung CT scans. Standard computer vision techniques were previously used to reduce noise, improve the quality of images, create handcrafted features, or test out traditional machine learning algorithms like SVM or K-means.

Deep learning techniques have greatly enhanced computer vision's capabilities recently, and the field has become increasingly important in the medical and healthcare sectors. Although deep learning has shown promising results in various applications, there are still challenges that need to be addressed. These challenges include the need for few-shot learning techniques and the lack of annotated data and ground truth.

9.7 SUMMARY

The use of computer vision technology will rise as AI is more pervasively incorporated into our daily lives. The need for professionals knowledgeable in computer vision systems will increase as it becomes more prevalent in our culture.

Computer vision engineers must keep up with the industry's most recent trends and developments to stay ahead of the curve and take advantage of recent breakthroughs. It is important to note that in 2022, there is a growing trend of using Transformers in computer vision tasks, mobile-focused deep learning libraries, and PyTorch Lighting.

In addition, edge devices are getting stronger, and companies are looking toward mobile versions of their goods and services. Deep learning libraries and mobile-centric packages are worth monitoring, since their use is anticipated to rise in the upcoming year.

Expect AutoML capabilities to increase in 2022, along with the expansion of machine learning frameworks and libraries. As augmented and virtual reality applications advance, CV engineers will be able to expand their expertise into new fields, including creating simple, effective ways to replicate real items in 3D space. Computer vision applications will influence future change and influence, and the infrastructure technology that supports computer vision systems will continue to advance.

REFERENCES

1. Velliangiri, S., Karthikeyan, P., Joseph, Iwin Thanakumar and Kumar, Satish AP. (2019). Investigation of Deep Learning Schemes in Medical Application. In 2019 International Conference on Computational Intelligence and Knowledge Economy (ICCIKE), pp. 87–92. IEEE.
2. Cosido, O. et al. (2014). Hybridization of Convergent Photogrammetry, Computer Vision, and Artificial Intelligence for Digital Documentation of Cultural Heritage – A Case Study: The Magdalena Palace. In 2014 International Conference on Cyberworlds, pp. 369–376. https://doi.org/10.1109/CW.2014.58.
3. Shelhamer, E., Long, J. and Darrell, T. (2016). Fully Convolutional Networks for Semantic Segmentation, ArXiv160506211 Cs. Accessed: May 08, 2022. [Online]. Available: http://arxiv.org/abs/1605.06211
4. Babatunde, O. H., Armstrong, L., Leng, J. and Diepeveen, D. (2015). A Survey of Computer-Based Vision Systems for Automatic Identification of Plant Species, J. Agric. Inform., vol. 6, no. 1, Jan. 2015, https://doi.org/10.17700/jai.2015.6.1.152.
5. Rautaray, S. S. and Agrawal, A. (2015). Vision-Based Hand Gesture Recognition for Human-Computer Interaction: A Survey, Artif. Intell. Rev., vol. 43, no. 1, pp. 1–54, Jan. 2015, doi: 10.1007/s10462- 012-9356-9

6. Joshua, J. (2017). Information Bodies: Computational Anxiety in Neal Stephenson's Snow Crash. Interdisciplinary Literary Studies, 19(1), 17–47.

7. North of 41 (2020). What is the difference between AR/MR/VR/XR? Access Date: 15/12/2021 https://medium.com/@northof41/whatreally-is-the-difference-between-ar-mr-vr-xr-35bed1da1a4e.

8. Balaji, R., Deepajothi, S., Prabaharan, G., Daniya, T., Karthikeyan, P. and Velliangiri, S. (2022). April. Survey on Intrusions Detection System using Deep Learning in IoT Environment. In 2022 International Conference on Sustainable Computing and Data Communication Systems (ICSCDS) (pp. 195–199). IEEE.

9. Thomas, P. C. and David, W. M. (1992, January). Augmented reality: An Application of Heads-up Display Technology to Manual Manufacturing Processes. In Hawaii International Conference on System Sciences (pp. 659–669).

10. Karthikeyan, P. (2021). An Efficient Load Balancing Using Seven Stone Game Optimizations in Cloud Computing. Software: Practice and Experience, 51(6), 1242–1258.

11. Huang, T. S. (1996). Computer Vision: Evolution and Promise. CERN European Organization for Nuclear Research-Reports-CERN (pp. 21–26).

12. Russakovsky, O., Deng, J., Su, H., Krause, J., Satheesh, S., Ma, S., Huang, Z., Karpathy, A., Khosla, A., Bernstein, M. and Berg, A.C., 2015. ImageNet large scale visual recognition challenge. International journal of computer vision, 115, 211–252.

13. Saad M. Almutairi, S. Manimurugan, Majed M. Aborokbah, C. Narmatha, Subramaniam Ganesan, P. Karthikeyan, "An Efficient USE-Net Deep Learning Model for Cancer Detection", International Journal of Intelligent Systems, vol. 2023, Article ID 8509433, 14 pages, 2023. https://doi.org/10.1155/2023/8509433

14. Barioni, R. R., Figueiredo, L., Cunha, K. and Teichrieb, V. (2018, October). Human Pose Tracking from RGB Inputs. In 2018 20th Symposium on Virtual and Augmented Reality (SVR) (pp. 176–182). IEEE.

15. Sarveshwaran, Velliangiri, Joseph, Iwin Thankumar, Maravarman, M. and Karthikeyan, P. (2022). Investigation on Human Activity Recognition Using Deep Learning. Procedia Computer Science, 204, 73–80.

16. Davies, Marion (2021). Pros and Cons of the Metaverse. Access Date: 15/12/2021. https://www.konsyse.com/articles/pros-and-cons-of-themetaverse/.

17. Mehul (2020). Object Tracking in Videos: Introduction and Common Techniques. Access Date:15/12/2021. https://aidetic.in/blog/2020/10/05/object-tracking-in-videos-introductionand-common-techniques/.

18. Rajan, D.P., Premalatha, J., Velliangiri, S. and Karthikeyan, P. (2022). Blockchain-Enabled Joint Trust (MF-WWO-WO) Algorithm for Clustered-Based Energy Efficient Routing Protocol in Wireless Sensor Network. Transactions on Emerging Telecommunications Technologies, pp. 1–17.

19. Lee, L. H., Braud, T., Zhou, P., Wang, L., Xu, D., Lin, Z. and Hui, P. (2021). All One Needs to Know About Metaverse: A Complete Survey on Technological Singularity, Virtual Ecosystem, and Research Agenda. arXiv preprint arXiv:2110.05352.

20. Bharathy, A.V. and Karthikeyan, P. (2022). Security and Privacy Policies in Artificially Intelligent 6G Networks: Risks and Challenges. In Challenges and Risks Involved in Deploying 6G and NextGen Networks (pp. 1–14). IGI Global.

21. Jain, Ayush, Gandhi, Kaveen, Ginoria, Dhruv Kumar and Karthikeyan, P. (2021). Human Activity Recognition with Videos Using Deep Learning. In 2021 International Conference on Forensics, Analytics, Big Data, Security (FABS), vol. 1, pp. 1–5. IEEE.

22. Rajagopal, R., Karthikeyan, P., Menaka, E., Karunakaran, V. and Pon, H. (2023). Disease Analysis and Prediction Using Digital Twins and Big Data Analytics. In New Approaches to Data Analytics and Internet of Things Through Digital Twin (pp. 98–114). IGI Global.

10 Study of Computer Vision Applications in Healthcare Industry 4.0

Mahesh Lokhande
Kalpanadevi D
Vandana Kate
Arpan Kumar Tripathi
Prakash Bethapudi

CONTENTS

DOI: 10.1201/9781003345411-10

10.1   INTRODUCTION

The relationship between robots and humans is being revolutionized by computer vision (CV), a relatively new artificial intelligence (AI) technology. The main goal of this research is to develop intelligent computer algorithms that can comprehend and analyze visual input without having to be explicitly programmed in a certain language or program. As long as we only rely on our own senses, doctors are responsible for inspecting and diagnosing illnesses and other health problems. What they can perceive with their eyes, ears, and touch will determine the outcome. Perception can only be perceived by how human sight and intellect see reality. In contrast, CV is what enables machines to perceive, examine, and comprehend visual information. Additionally, CV performs better the more data there is. According to researchers, images make up to 90% of all input data used in healthcare. It creates a wide range of possibilities for enhancing patient care and the effectiveness of the healthcare sector as a whole through the training of computer vision algorithms. To put it another way, automating procedures that rely on picture identification can improve treatment while requiring less input from people. This is a top priority since it will free up medical personnel to concentrate on more complicated issues. Visual analytics have been used extensively to enhance human capacities. Computer vision technology helps in the healthcare department by enabling healthcare professionals to precisely monitor the illness or condition of a patient by reducing incorrect treatment or inaccurate diagnoses [1, 2].

10.2   GAUSS SURGICAL

Gauss Surgical has been able to leverage computer vision technology to develop a powerful blood-monitoring system that can estimate blood loss in real time. The computer vision-driven solution is also more capable of determining the hemorrhage conditions in patients, as well as maximizing the transfusions. Gauss Surgical's Triton system captures blood images on sponges and suction canisters using an iPad-based app. This application is based on computer vision and machine language algorithms. Healthcare professionals in hospitals use this application to calculate the amount of blood loss during Cesarean deliveries or surgical operations.

In 2017, Gauss Surgical performed an analysis on the amount of blood on the suction canister or surgical sponges collected from the cesarean deliveries by using this application and found that the result is more accurate than obstetricians' visual estimates. The anesthesiologist uses various factors, including the patient's hemodynamic stability, cardiac status, and levels of hemoglobin (Hb) and hematocrit (Hct), along with the estimated blood loss (EBL), to determine the need for intraoperative blood product transfusions. Hb and Hct stage interpretation is confounded by means of intravascular hemodilution or hemoconcentration ensuing from the infusion of crystalloid or colloid solutions, blood product transfusion, 0.33 spacing, or acute hemorrhage into the surgical field. Blood loss estimates are normally visible estimates, and as such are challenging to standardize.

Having an accurate, real-time measure of blood loss, in addition to laboratory measures of Hb/Hct and clinical judgment based on the patient's hemodynamic stability and cardiac status, could significantly improve the decision-making process around transfusion, including whether or not to transfuse and how much blood to transfuse [3]. Estimates of blood loss during surgery have mostly been based on ocular assessments agreed upon by the surgeon and anesthesiologist. Estimates of blood loss have been demonstrated to be extremely wrong, with doctors frequently underestimating high blood loss volumes and overestimating low blood loss volumes, which almost certainly leads to under- or over-transfusion. Although simulations and didactic training to enhance clinicians' abilities to stimulate blood loss have been suggested, it has been demonstrated that these abilities degrade over time. Additionally, there is no correlation between experience level and the precision of providers' estimates. Although this method is impractical for real-time intraoperative use and highly sensitive to the presence of confounding non-sanguineous fluids (e.g., saline, ascites, amniotic fluid) on absorbent media, it has been investigated as a method of gravimetric estimation of blood loss by weighing soaked laparotomy sponges and subtracting their known dry weight. In research investigations, methods for washing and measuring Hb content from blood-absorbing media have been referred to as standards for the estimation of intraoperative blood loss, but they are also unsuitable for use in real-time during surgery.

In order to accurately estimate the Hb mass (mHb) absorbed by surgical sponges, the Triton system integrates mobile imagery with computer vision and machine learning algorithms. The Triton system enables intraoperative scanning of surgical sponges by the circulating nurse when they are removed from the sterile surgical field as part of the regular procedure of documenting sponge count using a camera-enabled mobile app native to the iPad 2 (Apple Inc., Cupertino, CA) [4, 5].

The mobile application simultaneously captures photographs of blood-soaked sponges, encrypts them, and wirelessly sends them to a distant server using the secure HTTPS protocol. The server analyzes relevant images and parses photographic and geometric data using Gauss feature extraction technology (FET). The server uses proprietary classifiers and computational models to extract photographic and geometric data from pertinent regions of interest and compute mHb utilizing Gauss FET. To filter out the impacts of extraneous non-sanguineous fluids (such saline) and to account for variations in intraoperative illumination conditions, FET is built using fine-grained detection, classification, and thresholding algorithms. Within seconds, the smartphone display receives a cumulative value of the mHb in sponges. This prospective, multicenter clinical study's goal was to evaluate the precision of the Triton system and FET in calculating the mHb lost onto laparotomy sponges during surgery. Figure 10.1 depicts the photographs demonstrating the sponge scanning and analysis process using the Triton system [6, 7].

10.2.1 STUDY POPULATION AND SURGICAL CASES

University of Texas MD Anderson Cancer Center (Houston, TX), Santa Clara Valley Medical Center (San Jose, CA), and Englewood Hospital and Medical Center (Englewood, NJ) have all given their approval for the research procedure (Englewood, NJ). At all sites, the need for written informed permission was forgone. Many bloody laparotomy sponges were predicted. Hence the study enrolled consecutive surgical

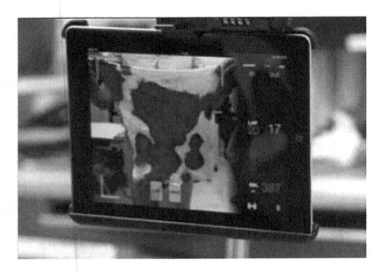

FIGURE 10.1 Photographs demonstrating the sponge scanning and analysis process using the Triton system.

patients. General surgery, obstetrics, orthopedics, and heart surgery were among the surgical procedures enrolled. For each procedure, there were no specified enrollment requirements. The patient's preoperative Hb/Hct measurements were recorded according to what was mentioned in the patient's medical file.

10.2.2 HEMOGLOBIN LOSS AND BLOOD LOSS MEASUREMENTS

When using and managing surgical sponges, the researchers adhered to their standard of care. Each instance ended with the collection of all laparotomy sponges for measurement using the Triton system, followed by rinsing and a Hb assay. After the data was collected, measurements were finished in two hours. The sponges were scanned using the Triton system with FET technology (Version 2.0.9, Gauss Surgical, Inc.), and the mHb loss as a result was measured (mHbTriton). The dry weight of the sponges was then subtracted from the weight of each case of sponges using a calibrated digital scale (A&D Co. Ltd., Tokyo, Japan) to get the total fluid weight, which is comparable to the gravimetric EBL (EBLWeight).

Then, using techniques identical to those employed in recent investigations, the Hb loss from each case was calculated by manually rinsing the red blood cell (RBC) content from the sponges (mHbRinse). 5,15 Sponges were rinsed either individually using a manual compression device or in batches using a centrifuge that was run at 1600 RPM for 45 seconds after being soaked in heparinized normal saline. After rinsing, a low-concentration hemoglobin analyzer (HemoCue Plasma/Low Hb, Hemocue AB, Ängelholm, Sweden) was used to determine the Hb concentration of the effluent solution (HbEffluent). A digital scale was used to measure the mass of the effluent solution and the volume of effluent (VEffluent) was calculated using the assumption that the mean fluid density was 1.0 g/ml. The mass of hemoglobin in the effluent was then determined using the formula below: mHbEffluent equals HbEffluent.

The recovery rates of the individual and batch rinse methods were independently characterized in a bench-top setting where banked blood was deposited on sponges in known quantities and mechanically extracted using the rinse methods, because it is not possible to recover 100% of the RBCs on sponges via rinsing. A linear regression analysis showed that batches of sponges (n = 11, 20 sponges/batch) had mean mHb recovery rates of 98.99% (95% CI 87.1% to 110.9%) while individual sponges (n = 116) had mean mHb recovery rates of 89.5% (95% CI 86.8% to 92.1%). The values of mHbEffluent in the clinical research were then adjusted using these characteristic recovery rates to generate a more accurate measurement of Hb loss on the sponge (mHbRinse).

10.3   CASE STUDY: VISION SYSTEM TRACEABILITY IN MEDICAL DEVICE MANUFACTURING APPLICATIONS

Industrial Vision Systems Ltd. (IVS, Oxford, UK; www.industrialvision.co.uk) purposefully incorporated XML support into the most recent version of IVS so that automated documentation could be produced when automated visual inspection processes were completed.. The applications division integrated the device created for Pfeiffer as a whole solution. The main inspection standard for the system was the separation of items based on established measurement standards to regulate the quality of the finished goods. The range of markets for machine vision solutions is expanding and becoming broader than ever.

The automotive industry still plays a significant role in the development of vision, but the pharmaceutical industry is opening up new application areas. The standard prerequisites for vision system installation, such as communication with the vision controller and Programmable Logic Controller (PLC) control of the vision controller, multitasking of the vision system, and documentation of the vision method, can be applied to various industries. The pharmaceutical business, more than any other where carefully specified check routine structures are required and documented, clearly has a greater requirement for this last point, especially within the limits of CFR 21/210/211/GAMP validation.

The IVS check routine's many inspection criteria make up the solution. The first inspection entails looking at the intermediate piston stroke length, which is a gauge of the pump's piston stroke and has a tolerance of ±0.01 mm. A check routine is created to differentiate between the five different pump types based on this result. The second inspection is looking at the translucent plastic body that houses the spray mechanism's spring and ball. Verification of these crucial elements is finished using a template matching function. Double ball bearings, bent springs, misaligned springs, and the incorrect intermediate piston are examples of common mistakes in this area. The key to the application is integrating the machine vision system into the cycle time of the machine. By optimizing the template matching function utilizing the special automatic wizard, it was possible to keep the image processing within the acceptable performance limits.

The entire approach included feeding the pump bodies onto a rotating plate; thus, the test position's positioning tolerance is well-stated. Two NCG212 1296 × 966 IVS Gigabit Ethernet digital cameras are mounted in a fixed position outside the carousel area. The PLC transmits a signal immediately to the I/O control in the IVS PC once the pump body has arrived at the inspection station on the carousel. The evaluation by both cameras is finished and sent back to the PLC within 100 msecs.

One I/O channel is used to individually indicate each feature failure. Excel is used to store information and to communicate data with the offices using an Ethernet connection. The pharmaceutical business has already emphasized the importance of documentation. Once a comprehensive solution has been established within IVS, the entire inspection criteria can be automatically stored as an XML document as illustrated. Since Pfeiffer needs this kind of traceability, this provides comprehensive information on each check function utilized and how it was configured. This makes the system as a whole very powerful when deploying many systems throughout the pharmaceutical business. With the understanding that the software has been specifically created to manage the intricate criteria demanded by the pharmaceutical business, the system is now consistently inspecting an annual production of 200 million pumps.

10.4   CASE STUDY: MEDICAL PARTS GET A CLINICAL CHECKUP

Vision-based inspection is used to examine multi-cavity injection molded subassemblies used in the medical industry.

Up until recently, inspecting injection-molded medical equipment required skilled personnel to visually examine each item for consistency and quality. This was a time-consuming process. Each item had to be scrutinized during the inspection of one such equipment, a central barrel for a drug injection system, to make sure it complied with a stringent set of requirements before being approved for shipment, placed aside for recycling, or rejected. Industrial Vision Systems, Ltd. has created a machine vision system to automate this operation and do this inspection at a rate of over 100 pieces per minute. The system's devices consist of a black injection-molded shaft with a central depression into which a yellow plastic band has been molded. When the gadget is in use, the center barrel moves down a cylindrical tube to eject the medication, and the yellow-molded plastic band serves as a visual cue that the right dosage has been provided.

A color-matching approach can then be used during the inspection phase to compare the bands in each of the four photos to see if the various portions of the image have the right RGB value and even coverage. Any discrepancies show that either the yellow band is completely missing or that there are dark places where it may have been molded wrongly. The check feature in the IVS program converts each RGB image into a gray-level image for this purpose. The degree of color similarity between the digitized images and similarly processed known reference colors is then indicated by the brightness of each pixel in the generated image.

10.4.1   PARALLEL PATTERNS

The degree of parallelism of the edges of the barrel's body is also determined by IVS software. Edge detection is employed to locate the locations in the image where measurements ought to be performed. Two lines are chosen from the generated image to stand in for the barrel's two edges. The placement of these outer edges and a pseudo-baseline are utilized to determine the angular offset and, consequently, the parallelism of the lines, because the parts are positioned horizontally in the tooling. The pixel data in an image can be used to detect shorts or flashes by analyzing the parallelism of lines along the edges of the object in the image. This is done by interpreting the location of individual pixels on the edges of the object. This is done by computing ideal straight lines from existing regions of interest in a picture using another function in IVS. To do this, the software looks at each individual pixel in the barrel edge photos and uses that information to generate reference lines, or places where pixels indicating edges should ideally be present.

The edges recognized by the program will produce a line that differs from the reference if shorts or flash are present on the outside of the body, offering a way to identify areas that may need to be reworked or rejected. The top of the body part was photographed to ensure that there were no flashes or shorts coming from the two molded flanges at the head of the part. To do this, edge detection is first used to identify the part's head from the image.

If the part is within acceptable parameters, specific locations of interest are measured and compared with a reference template. The program analyzes the device's features before deciding whether the part has to be reworked or rejected if it doesn't satisfy all the parameters that have been defined. The plastic injection-molded part undergoes tracking around the rotary plate, starting from the inspection station and continuing until it reaches one of the three other stations. At these stations, pneumatic actuators are used to remove the part from the plate and place it in the appropriate tote bins, ensuring that the individual devices are sorted correctly. Although the information from the image processing system can be utilized to quickly examine and classify the devices, it also has other advantages. The data can be used to estimate how any changes in allowable parameters would affect the yield of the product, because statistical data can also be gathered throughout the inspection process, after the machine has examined numerous batches of goods.

10.5   CASE STUDY: MEDICAL WELLS SCRUTINIZED BY VISION

In the past, workers manually checked trays of plastic wells used in microwell plates by taking them off a conveyor belt on a production line and visually inspecting them. The technique was vulnerable to human error due to the subjective character of the operation

To create a custom-built vision machine for a particular purpose, adhering to GAMP5 pharmaceutical regulations for full validation, a medical manufacturer in the UK sought assistance from engineers at IVS.

Ninety-six white plastic wells stacked in rows make up a typical microwell plate, which is used in labs as a miniature test tube to examine cell cultures. The wells are coated with active chemical compounds during a multistage production process. When serum or plasma is deposited in the wells during use, these substances bind to specific targets. When they do, they bring about color changes that are visible to medical equipment used to check for the presence of disease. The coated white plastic wells must maintain their color consistency so that when they are utilized, the color change recognized by the medical equipment gives a trustworthy indicator of the chemical changes occurring in the sample.

The problem is made worse by the fact that the coating process's chemicals themselves are colored. When serum or blood is tested for disease, the instrumentation could yield false findings, even though the chemicals should have been removed from the wells in the production process. Detecting any particle debris that might remain in the well as a result of the production process is equally important to maintain the accuracy of such tests as ensuring the consistency of the color of the wells, both indoors and outside. Four vision-based inspection stations built into the IVS system check the wells' interior and exterior colors and determine whether any particles are present. The system removes any wells that don't comply with the manufacturer's specifications from the tray, and then a last visual check is done to make sure the reject process was successful.

The coated plastic wells are examined at the first vision inspection station to ensure that their whiteness complies with a set standard. For this purpose, the station uses a Qioptiq Rodogon 105 mm f/5.6 lens on an L103 1K monochrome line scan camera from Basler (Ahrensburg, Germany; www.baslerweb.com). To protect it from ambient light, this camera is positioned inside a stainless steel shroud. The bottom of the tray is lit by a high-intensity MBRC red line light from Schott Moritex (Saitama, Japan; www.schottmoritex.com), which travels in time with the camera as it traverses the top of the tray of wells under servo control. One of the crucial features is the gray scale check performed by the vision system.

The monochrome line scan camera used for the assignment allows for the capture and processing of high-quality images in a reasonable one-second time frame, and the single wavelength red-line light ensures uniform illumination throughout all of the wells. One of the three Intel i7 industrial PCs in the system receives an image taken by the line-scan camera from above a tray of wells and transfers it for processing using NeuroCheck software from Neurocheck GmbH. (Remseck, Germany; www.neurocheck.com). The gray-scale intensity of the areas in the image bounded by the circular edge of the wells is then calculated sequentially by the NeuroCheck program.

In particular, it computes the average, the standard deviation from the average, and the greatest and minimum gray-scale intensities. Then, the gray-scale intensities are contrasted with a reference set of allowable gray-scale values. In order to teach the system what range of gray-scale values is acceptable, this template is built during system setup by taking pictures of failed components. Should one or more of the wells fail the inspection, the PLC logs the information after receiving the location of the failing well via an Industrial Ethernet/IP network.

10.5.1   COLOR CHECK

The tray of wells is indexed by the walking beam system to the second inspection station once the inside of the wells has been examined for gray-level intensity. Here, the wells are examined to see if any colorful chemicals exist and to make sure there isn't any particle stuff inside of them. This vision examination station does so by utilizing two NeuroCheck NCF120c area-scan cameras equipped with 25 mm f1.4-22 Ricoh Pentax lenses (Tokyo, Japan; www.ricoh-imaging.co.jp).

The cameras are enclosed in a stainless-steel shroud, and an on-axis flat dome, light from IMAC Machine Vision Lighting is utilized to illuminate the top of the tray of wells (Moriyama, Japan; www.kkimac.jp).

This light provides the needed wide angle of lighting while taking up a tiny fraction of the space of conventional dome lights. For processing, the two area scan cameras' collected images are sent across a FireWire interface to a second PC powered by an Intel i7 processor. The same regions of interest in the images that were examined during the gray-scale analysis procedure, specifically, the area inside the wells, are examined by NeuroCheck software.

At this stage, however, the images are checked for color consistency. To do so, the system is first calibrated by presenting the system with a number of rejected parts. RGB values of those images are then used to create a template of color values. During operation, the software compares the RGB values of the images of the wells with the range of template color values to determine whether a product falls within a given tolerance range. At this stage, the interior of the wells is also checked to determine if any particulate matter has been left in the well during production. To do so, the NeuroCheck software performs a thresholding operation on the image to separate the image background with regions of the image that correspond to any foreign objects. If a well fails due to its unacceptable color, or because of the presence of particulates, the location of the well in the tray is transferred over the Profibus network to the PLC.

The third stage of vision inspection, where the wells' colors are examined, is then indexed to the tray of wells. To do this, a picture of the bottom of the wells is taken using a Basler L304KC color line scan camera equipped with an Opto Engineering TC16M080–88.4 mm telecentric lens (Mantova, Italy; www.opto-engineering.com). A fiberoptic light from below that moves in time with the camera and a backlight from above both shine on the underside of the tray as the camera travels across it.

A third Intel i7 industrial PC receives the image of the well bottoms via a camera link interface. The NeuroCheck software in this case recognizes the regions in the photographs that correspond to the bottom of the wells and contrasts the color in the images with a second template that represents a spectrum of reference colors in RGB color space. As before, the PLC receives control of the well's location within the tray if any well fails the inspection procedure.

10.6   CASE STUDY: COMPUTER VISION FOR MONITORING TUMORS USING IMAGE SEGMENTATION

10.6.1   MONITORING TUMORS IN THE LIVER

The Amsterdam University Medical Centers (AUMC, Amsterdam, the Netherlands) is included in one of the most popular computer vision case studies. In addition to offering patient treatment, AUMC does scientific research and functions as a teaching hospital. Employing digital imaging and communications in medicine (DICOM) pictures from CT scans, they are using SAS computer vision to track the development of liver cancers.

The third most prevalent cancer in the world, colorectal cancer regrettably progresses to the liver in nearly half of patients. Colorectal cancer is the third most prevalent cancer in the world, and unfortunately, it progresses to the liver in nearly half of patients. Patients with large tumors may need to receive chemotherapy to reduce the tumor size before being evaluated as candidates for surgical excision [8, 9].

- After a course of treatment, radiologists use CT scans to manually assess the tumors. Radiologists can then determine whether a tumor has decreased or if its appearance has changed. The outcomes of this assessment define the subsequent steps in a patient's treatment plan, which may include surgery, the use of a different chemotherapy regimen (drugs, dosages, and timing), or other options.
- Radiologists face numerous difficulties as a result of this time-consuming approach.
- For radiologists, evaluating tumors is a time-consuming process. Additionally, on each CT scan, only the two largest tumours are routinely measured, potentially omitting data from any smaller tumors.

Additionally, subjectivity in the manual assessment makes it possible for diverse assessments to be made by a group of radiologists. The experiment began by using data from 52 cancer patients to create a deep learning model. 1,380 metastases were segmented, and every pixel was examined. This trained the system to recognize tumor traits right away. Image segmentation, which is utilized in computer vision, was the deep learning model that was used. For easy analysis, an image can be divided into several segments using image segmentation. To manually recognize the parts without computer vision would be exceedingly challenging, especially when the contrast is low between the lesion and the organ [10, 11].

10.6.2 IMAGE ACQUISITION

The DICOM pictures and DICOM-RT contours were first loaded. These include contour maps of the tumors and the liver as well as anatomical photos of the liver. Be aware that the contour maps are radiation therapy-related DICOM RT images shown in Figures 10.2 and 10.3. For image segmentation, both the images and the contours are required.

10.6.3 ANALYZING THE IMAGE

What matters is how big the tumor actually is. The three-dimensional (3D) volume of the tumor must be calculated in order to quantify the mask image in order to provide an answer. To convey this information to the doctors, the computation results might subsequently be plotted. The tumor volumes are displayed in the graph (Figure 10.3) at various stages of treatment.

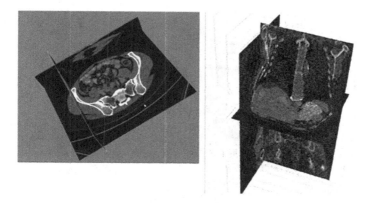

FIGURE 10.2   DICOM image with liver lesion surfaces.

10.6.4 APPLYING THE INSIGHTS

After the deep learning model is trained, it can then be deployed to evaluate new images. This analysis was done using a Jupyter notebook in combination with the biomedical (BioMedImage) action sets from SAS Viya and SAS Deep Learning Python (DLPy).

10.7   CASE STUDY: DETECTION OF CANCER WITH COMPUTER VISION

Computer vision has had a significant impact on the field of life sciences. It has enabled the development of more accurate diagnoses, treatments, and research methods, ranging from early cancer detection to improving plant health.

10.7.1 IDENTIFYING NEUTROPHILS WITH COMPUTER VISION

Neutrophils are a type of white blood cell that play a key role in animal immune systems. In humans, neutrophils help heal injuries and fight infections. Identifying and

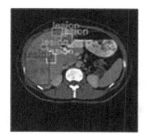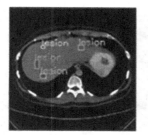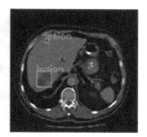

FIGURE 10.3   Bounding boxes around the lesions.

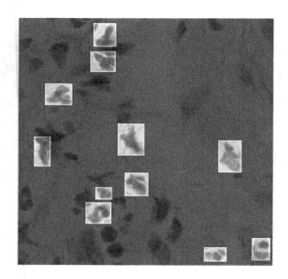

FIGURE 10.4   Detection of cancer.

measuring their presence in microscopy is an essential step in assessing the efficacy of given experiments. Figure 10.4 depicts the detections of cancer in the image.

Mateo Sokac points out that dry and immersion microscopy are the two methods his labs use to identify neutrophils. Immersion is typically more accurate, slower, and more costly. Dry microscopy, on the other hand, is quicker, less expensive, and of inferior quality. To determine which aspects of a cell population are enriched in animal populations, Mateo's team is now conducting tests. These similar methods will eventually be used on people to advance the study of cancer (Figure 10.5). Mateo aimed to train a model to more precisely identify neutrophils in dry microscopy as accurately as one can in immersion in order to improve the efficiency of his lab. Mateo trained a model using images labeled from immersion images, but only using dry microscopy for image validation. Mateo had 127 photos at first, which Roboflow increased to 451, and he saw a mean average precision of 0.70, up 10% from his unaugmented results. Mateo easily attempted training multiple architectures via the

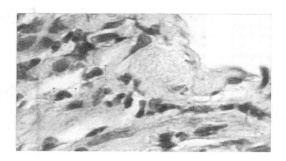

FIGURE 10.5   Individually labeled neutrophils from Mateo's dataset.

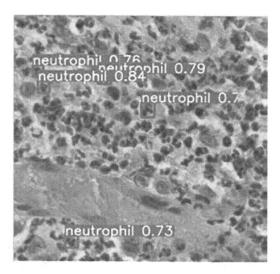

FIGURE 10.6 Predictions for neutrophil identification on an outputted image tile.

Roboflow Model Library, including YOLOv5, EfficientDet, MobileNetSSD, and more. Using Image Tiling to Improve Performance Moreover, Mateo's team uses tiling to enhance model performance because the input images are so enormous (27,000 by 27,000 pixels). In order to train a model and make performance inferences on images with increased pixel densities, tiling is the method of splitting a large input image into smaller input images. Breaking an image into four sections allows the input image to keep its original image quality, for instance, if the input image is 1600 × 1600 but a model can only perform on, say, 800 × 800 input images. Easy tiling is supported by Roboflow Pro as a preprocessing step.

After an image is tiled, it is run through the model to count neutrophils, and the team then reconstructs the individual tiles to create a full image output. The numbers refer to the model's predicted confidence of each label. A prediction for one of these individual tiles is depicted in Figure 10.6.

10.8 CASE STUDY: COMPUTER VISION FOR PREDICTIVE ANALYTICS AND THERAPY

Recent advancements in rapid prototyping and 3D modeling technologies have greatly facilitated the development of medical imaging modalities such as CT and MRI, offering considerable promise for the use of computer vision technology in surgery and the treatment of various diseases. These imaging modalities, combined with techniques such as DTI tractography and image segmentation, allow for more precise diagnosis and treatment planning..

Gargiulo et al. in Iceland "New Directions in 3D Medical Modeling: 3D-Printing Anatomy and Functions in Neurosurgical Planning" 3D-model the skull base, the tumor, and five eloquent fiber pathways. The authors offer a fantastic possible therapeutic strategy for sophisticated neurosurgical training [12].

The elderly are vulnerable to falls, which can hurt their bodies and have detrimental psychological effects. T.-H. Lin and colleagues in Taiwan have developed an interesting line-laser obstacle detection system for "Fall Prevention Shoes Using Camera-Based Line-Laser Obstacle Detection System" to help prevent falls in the elderly [13]. The system uses a laser line that travels along a horizontal plane and is positioned at a set height above the ground. Because the optical axis of a camera is positioned at a specific angle to the plane, the camera can view the laser pattern and identify potential impediments. Unfortunately, the system's architecture makes it more suitable for indoor than outdoor uses.

One of the many computer vision issues being extensively researched is human activity recognition (HAR). Human activity recognition in healthcare involves using sensors and machine learning algorithms to monitor and analyze physical activities and movements of individuals in healthcare settings. The technology can be used to assist in the monitoring of elderly or disabled patients, identify abnormal behavior or activity patterns, and provide insights for personalized care plans. This can help healthcare providers make informed decisions and improve patient outcomes. They classify and define human activities in accordance with these divisions into three levels: action primitives, actions/activities, and interactions [14, 15].

10.8.1 RESULTING IMAGES PICTURES

After conducting inference on each individual tile, the tiles are rebuilt into their final $27,000 \times 27,000$ pixel shape. The outcomes allow Mateo's team to assess experiment efficacy more accurately and rapidly. The final shots are absolutely beautiful.

After the model has identified the outputted neutrophil, it may become evident that tiling is necessary to capture the detailed granularity required for Mateo's team to achieve accurate results when examining the full-sized image. Figure 10.7 shows the fully reconstructed image from Mateo's lab.

FIGURE 10.7 A complete reconstruction of an image from Mateo's lab.

10.9   SUMMARY

CV has enormous potential in the medical industry by improving patient care. The use of CV might range from detecting early illness symptoms to monitoring the patient's well-being during and after the operation. By raising the standard of patient care, CV can perform wonders in the medical field. Essentially, CV problems include teaching computers to comprehend both digital images and visual information from the outside world. To do this, data from such sources may need to be gathered, processed, and analyzed in order to make decisions. The substantial formalization of complex problems into well-known, compelling problem statements is a defining feature of the development of machine vision.

REFERENCES

1. Anderson, P., He, X., Buehler, C., Teney, D., Johnson, M., Gould, S. and Zhang, L.: Bottom-up and top-down attention for image captioning and vqa. arXiv preprint arXiv:1707.07998 (2017).
2. Velliangiri, S., Karthikeyan, P., Joseph, I.T. and Kumar, S.A.: December. Investigation of deep learning schemes in medical application. In: 2019 International Conference on Computational Intelligence and Knowledge Economy (ICCIKE), (pp. 87–92). IEEE (2019).
3. Bai, W., Sinclair, M., Tarroni, G., Oktay, O., Rajchl, M., Vaillant, G., Lee, A.M., Aung, N., Lukaschuk, E. and Sanghvi, M.M., et al.: Human-level CMR image analysis with deep fully convolutional networks. arXiv preprint arXiv:1710.09289 (2017).
4. Cai, J., Lu, L., Xie, Y., Xing, F. and Yang, L.: Improving deep pancreas segmentation in CT and MRI images via recurrent neural contextual learning and direct loss function. In: MICCAI, Springer (2017).
5. Cerrolaza, J.J., Summers, R.M. and Linguraru, M.G.: Soft multi-organ shape models via generalized PCA: A general framework. In: MICCAI. pp. 219–228. Springer (2016).
6. Gibson, E., Giganti, F., Hu, Y., Bonmati, E., Bandula, S., Gurusamy, K., Davidson, B.R., Pereira, S.P., Clarkson, M.J. and Barratt, D.C.: Towards image-guided pancreas and biliary endoscopy: Automatic multiorgan segmentation on abdominal CT with dense dilated networks. In: MICCAI. pp. 728–736. Springer (2017).
7. Greff, K., Srivastava, R.K. and Schmidhuber, J.: Highway and residual networks learn unrolled iterative estimation. arXiv preprint arXiv:1612.07771 (2016).
8. Heinrich, M.P., Blendowski, M. and Oktay, O.: TernaryNet: Faster deep model inference without GPUs for medical 3D segmentation using sparse and binary convolutions. arXiv preprint arXiv:1801.09449 (2018).
9. Heinrich, M.P. and Oktay, O.: BRIEFnet: Deep pancreas segmentation using binary sparse convolutions. In: MICCAI. pp. 329–337. Springer (2017).
10. Hu, J., Shen, L. and Sun, G.: Squeeze-and-excitation networks. arXiv:1709.01507 (2017).
11. Jetley, S., Lord, N.A., Lee, N. and Torr, P.: Learn to pay attention. In: International Conference on Learning Representations (2018), https://openreview.net/forum?id=HyzbhfWRW.
12. Gargiulo, P., Árnadóttir, Í., Gíslason, M., Edmunds, K., Ólafsson, I. New directions in 3D medical modeling: 3D-printing anatomy and functions in neurosurgical planning. Journal of Healthcare Engineering, 2017;2017. doi: 10.1155/2017/1439643.

13. Lin, T.H., Yang, C.Y. and Shih, W.P., 2017. Fall prevention shoes using camera-based line-laser obstacle detection system. Journal of Healthcare Engineering, 2017; pp. 1–17.

14. Sevugan, A., Karthikeyan, P., Sarveshwaran, V. and Manoharan, R. Optimized navigation of mobile robots based on Faster R-CNN in wireless sensor network. International Journal of Sensors Wireless Communications and Control, 2022;12(6), pp. 440–448.

15. Jain, A., Gandhi, K., Ginoria, D.K. and Karthikeyan, P.: December. Human Activity Recognition with Videos Using Deep Learning. In 2021 International Conference on Forensics, Analytics, Big Data, Security (FABS), (Vol. 1, pp. 1–5). IEEE (2021).

Index

Printed in the United States
by Baker & Taylor Publisher Services